Bees
at the bottom
of the garden

Bees
at the bottom
of the garden

by
Alan Campion

Illustrated by Gay Hodgson

Bees at the bottom of the Garden was written in 1983, when I had little experience of beekeeping, in response to a request for a simple, clear and well illustrated guide book for new beekeepers. It was first published by A & C Black (Publishers) Ltd, in 1984, all copies were sold and it was re-issued in paperback in 1990. I believe that the fact that the re-issue also sold out can largely be attributed to the clear and technically helpful illustrations of Gay Hodgson, herself an experienced beekeeper. I have had frequent requests for copies of the book since it went out of print. I have also received many gratifying compliments about the ease of reference of the contents, including one or two from beekeepers who wished to purchase a new copy as their old one was coming apart or glued up with propolis! Compliments indeed.
Gay Hodgson and A & C Black (Publishers) Ltd have kindly given agreement for me to re-issue a revised edition.

Alan Campion 2000

Published by Alan Campion, 48 Hawthorn Road, Lincoln LN2 4QX

© Alan Campion 1984, 1990, 2001

First published by A & C Black 1984, re-issued in paperback 1990
Revised and republished by Alan Campion 2001
Trade Distribution: Northern Bee Books

Original design by Krystina Hewitt
Cover photograph by kind permission of John Phipps

ISBN 978-0-907908-97-5

A CIP catalogue record for this book is available from the British Library.

Printed and bound in China by 1010 Printing International Ltd.

Contents

The test of time

In the first edition of Bees at the Bottom of the Garden I attempted to write clearly about the generally accepted principles of beekeeping, avoiding, where possible, bias and controversy.

I received only one complaint which when investigated was the result of a misunderstanding, and easily explained to the satisfaction of the gentleman concerned. The principles have stood the test of time and still hold good. The arrival of Varroa in the UK has however necessitated a change in the schedule of Annual Hive management, the section on Braula and Varroa has been re-written and more advice on coping with Varroa is given on page 109.

Acknowledgements

I would like to thank all my friends who have each contributed to my enjoyment of beekeeping, and directly or indirectly helped in the compilation of the text of this re-write. I would especially like to thank Gay Hodgson for allowing us to again use her drawings, without which the book would be more difficult to understand and much less fun to use.

Introduction

Bees at the bottom of the garden has been written with three main aims in view. Firstly, to encourage the reader to 'think bees' and at least consider carefully how to add to the attractions of his garden by bringing in a colony of bees. Secondly, to offer the reader a ready-made management system when he has swallowed the bait (or nectar) and decided to take up the craft. And thirdly, to give my friend and fellow beekeeper, Gay Hodgson, an opportunity to put on public view her admirable drawings of bees and beekeeping topics.

At one time or another we must all have watched a honey bee on a warm summer day, going in a frantic way from one flower to the next. By her very industry she sets an example to us all. The idea of having a hive full of these industrious creatures winging out in every direction, returning heavy with nectar, and providing the beekeeper with a limitless supply of honey, with no effort on his part, must appear an attractive proposition. The true story, of course, is not exactly like this. The bees go diligently about their business when the weather allows, and at the end of the summer they might have stored away some honey. This honey would feed them through the long winter months, so the beekeeper who takes some for himself must leave the bees sufficient food to live on or they will all die of starvation. After a particularly wet summer when the bees are unable to forage freely the beekeeper might even have to prevent winter starvation by providing the bees with sugar syrup to supplement their stores.

One other aspect which appeals to beekeepers and non-beekeepers alike is the mystery and fascination of the honey bees themselves. In describing the 'mysteries of the hive' early writers were fascinated by the apparent order of things, the complete sharing of the necessities of life, and each bee working at its allotted task. Beekeepers who have kept bees for many years claim to learn something new every day, yet this very mystique, and the public's idea of the bee-master who alone understands and controls his bees can, I believe, discourage many would-be beekeepers. A beekeeper in the United Kingdom is still a comparative rarity and yet in some countries it is as usual to keep a hive of bees in the garden as to grow tomatoes.

I have attempted to *describe* phenomena rather than explain them. For example, a foraging bee will visit many flowers over a wide area to collect her load of nectar. Having collected it she will fly in the well-known bee-line straight from the last collection point to her hive. The serious beekeeper must look to other books for theories to explain this; personally I am prepared to be full of admiration for the achievement and leave it at that.

In no way does this book set out to tell you all there is to know about bees and beekeeping — with less than ten years' experience of the craft at the time of writing it would be presumptuous of me to attempt to do so. However, I believe that my relative inexperience and the fresh memories of my many mistakes can be put to positive advantage. I well recall my panic the first time our bees swarmed, and the problems involved in following instructions from a book whose writer assumed every beekeeper had an inexhaustible supply of equipment, or unlimited funds to go out and buy it!

While on the subject of money, I have purposely not quoted prices of equipment since these can change before the ink dries. The reader can obtain price lists and catalogues by contacting the manufacturers listed in the appendix.

The management system described here is no new breakthrough in beekeeping. It is practised, with variations, by many beekeepers with a variety of bees, hives and apiary sites. I have merely attempted to set out in the simplest way a system requiring the minimum involvement and effort from the beekeeper and to suggest ways in which his hobby can be rewarding, free of apprehension, not over-expensive and which will hopefully repay the beekeeper with a surplus of honey for his own use.

Unlike dogs, cats and many farm animals kept by man, bees do not require constant daily attention; during the winter months they need very little attention at all. At different times, however, the beekeeper must examine and handle his bees, which does take up some time. Before letting yourself in for something you might later regret, you may care to read 'The beekeeper's year' (p. 92) and decide if you can spare the bees your time!

Minimum requirements

Obviously the bees themselves In most cases this involves buying the bees on frames from another beekeeper or supplier of beekeeping equipment. In May, June or July it is often possible to obtain your bees for nothing by collecting a swarm. This usually means a one-year wait before any surplus honey can be taken, but it may appeal to those really trying to take up beekeeping 'on the cheap'. Being the hub of the whole business, bees are dealt with separately and in greater depth in the next section.

A hive Your bees will need a home to live in. There is a wide range of hives available but in practice the majority of beekeepers in any particular area use the same types of hive. You, as the beekeeper, are advised to do the same. Hives, too, are dealt with separately in a later section.

A site on which to keep your bees — the apiary This need be no more than an area the size of a dining table. The bottom of the garden is a good rule of thumb unless this position causes annoyance to neighbours. There is more later on the siting of hives.

Protective clothing The well-dressed beekeeper wears a veil, a bee suit, gloves and wellington boots. It is surprising how much better angry bees look from inside this protection. Some items you may have to buy but beekeepers, with few exceptions, are not fashion conscious and 'anything goes' as long as you are inside and the bees stay out.

Spare hive and equipment Happiness in beekeeping is having a spare hive and equipment to deal with a colony which has decided to swarm. The spare equipment can be second-hand, second-rate, home-made or borrowed. It will normally only be in use for a few weeks, but it is vital.

Catalogues of beekeeping equipment There are several good suppliers of beekeeping equipment. The major British suppliers are listed in the Appendix. Most will send a catalogue of their products on request. In most cases this is also a useful guide to the use and management of the bees and equipment.

A beekeeping friend Many beekeepers are well-known as such in their district and you may

possibly have a contact already. If not, enquire about beekeeping associations at local newspaper offices, information centres, libraries or even the local police station where lists of beekeepers prepared to collect swarms are often kept! Read this book before making any contact and you will be better able to speak the beekeeper's language. He will doubtless disagree with much that I think is important, but just listen and ask questions and you will soon have a friend of inestimable value in your new hobby.

Honey extracting equipment Borrow this, at least initially, from your new friend or take it on loan, when required, from a local beekeeping association.

Add to this a smoker and a hive tool and you have the minimum requirements for your venture into the new world of beekeeping.

The honey bees kept by beekeepers in Britain and in many other countries all belong to the species known as Apis mellifera. For millions of years before man arrived on the scene honey bees lived and thrived in forests around the world, building their comb and making their homes in hollow trees. In different geographical areas the bees developed different characteristics; these differences were mainly of colour and size but also often of behaviour.

A hundred years ago the British Black bee held sway in this country. It was said to be tough, hardworking, thrifty and was of course black in colour. On the other side of the Channel the French had a similarly coloured bee which was also hard-working but which was apparently prone to bad temper and sudden attack. The Dutch meanwhile had their own brown bees; the Austrians had a light-coloured bee with white stripes in the form of a Carniolan, which was gentle but prone to swarming; the Russians kept a type of Caucasian honey bee which was grey, with white bands, and gentle; while the Italians developed a bee which was distinctively light brown in colour, and ideally suited to the warm Italian climate.

Then, at about the same time that Britain was experiencing the horrors of the First World War, a destructive force in the form of the disease now known as Acarine appeared and all but wiped out the indigenous British Black bee. After the War a restocking scheme was introduced and bees were brought over in large quantities from Holland, France and other countries. Because a virgin honey bee queen mates with more than one drone and does not limit herself to the drones from her own hive it was inevitable that most of the few colonies of pure British Black bees which survived Acarine would become crossed, to a greater or lesser extent, with bees from all over the Continent, so that the resulting characteristics of the bees varied widely from colony to colony.

Some of these characteristics were considered undesirable by the beekeepers of the time and so, with the benefits of modern means of transportation, honey bees were brought in from overseas to alter or improve them. This search for a better bee has gone on ever since, with bees being imported from many parts of the world.

The Italian bee especially was considered highly desirable and imported into Britain, as well as to many other parts of the New World. It is distinctive in appearance, being orange brown in colour, with the workers having three yellow brown bands on their darker coloured abdomens. The other important and valued characteristics of the Italian bee are its docile temper, reluctance to swarm, apparent resistance to disease and prolific breeding pattern.

Italian queen bees were available in quantity from May onwards and many beekeepers purchased imported Italian queens to replace old queens, improve their bees, and control swarming for that year, all at the same time. This has resulted in the yellow brown banding of the Italian bee turning up in many colonies where most bees are of a darker colour. The disadvantage of the pure Italian bee in this country is its tendency to rear brood to suit a warm Italian-type summer. In a warm climate the resulting strong colony will harvest a tremendous crop of honey, but in a wet British summer the Italian bees can literally starve. To keep Italian bees pure a regular supply of queens from Italy was required. However in the early eighties a bee pest called Varroa appeared in parts of Europe, was first found in Southern England in 1992, and is now considered to be present throughout most of the United Kingdom. This resulted in the British Government restricting the movement of bees into the country, so that most Italian bees found in this country are probably less pure than their ancestors.

The net result of all this is that the majority of bees in Britain are a mixture of some or all of the types mentioned, with a mixture of colours and characteristics, some good, some bad. Within a few generations the bees in an area do, however, become accustomed to its climate and vegetation, and you are strongly recommended to bear this in mind when obtaining your bees. Bees from the north are less likely to thrive in the south than those already acclimatised to the south, and vice versa. So the first suggestion for a would-be beekeeper is to find a beekeeping friend and make enquiries about purchasing bees from your area. The other characteristics you should be looking for are good nature, hardiness, a reluctance to swarm and an

Worker bee

inclination to gather lots of honey. Few beekeepers will be prepared to guarantee all these, but good temper is probably the most important. If you arrange to buy bees locally you can probably handle the bees first to test this factor, and you have at least some after-sales service if things go wrong. Other useful advice for the would-be beekeeper is to delay the purchase of your own bees until you have met more than one beekeeper and looked at *several* colonies of bees.

There are various diseases which afflict bees and to ensure against purchasing diseased bees it is a wise precaution to have two tests done on any colony you propose to buy. The first is to ask the seller to arrange for the colony to be inspected by the Bees Officer from the Ministry of Agriculture, Fisheries and Food. There should be one for your area and he should inspect each colony every few years to ensure no brood diseases are present. The local MAFF, or beekeeping association, will help in making contact, but a prompt inspection depends on locality. The second test is to fill a matchbox with bees and send them for analysis. This is usually carried out by the County Beekeeping Instructor in those counties fortunate enough to have one, or by the National Beekeeping Centre at Luddington (see p. 110). Any diseases detected will be reported back to the sender. Buying locally the seller can sometimes be persuaded to assist in siting the hive and settling the bees in the chosen site, but many sellers might consider the tests for diseases to be either superfluous or insulting!

Another completely different method is to purchase your bees, probably in the form of a nucleus, from a reputable supplier of bees and beekeeping equipment. Such suppliers have their reputations to maintain and are unlikely to deter a potential regular customer by supplying diseased or bad-tempered bees.

The cheapest way of all to start beekeeping is to be ready with your hive and equipment in the hope of obtaining a swarm for nothing. I don't recommend it for a beginner. Most of the swarms I have taken have been disappointing. There is usually no honey surplus the year the bees are taken, they can bring in disease, they usually swarm again the next year, and they often need to be requeened. But perhaps I have been unlucky.

Next to the cost of the bees themselves the hive is almost certainly the second most expensive item required to start beekeeping, so care should be taken to make sure that the money is wisely spent and the most suitable hive is purchased.

Hives do not occur naturally, they are a comparatively recent arrival on the bee scene. Over the ages honey bees have developed the ability to search out and find hollow cavities, usually in trees, move in *en masse*, and set up home. Inside the tree the worker bees produce beeswax and fashion it into the honeycomb, which they then use as a living area for storage of nectar and pollen, and for the rearing of their brood, the developing bees. Under natural conditions this honeycomb will be suspended from the roof of the cavity and will be as rigidly fixed as possible. The comb will normally be fashioned in slabs with the openings into each cell facing the openings of the cells on adjacent combs. Between each comb the bees leave a space wide enough for them to move about and have access to each cell opening.

Earlier beekeepers, observing that the requirements for a good bee home were thus a hollow darkened area of sufficient size to accommodate a full colony, protection against the weather (especially against rain), and an entrance which could be defended by the bees against marauders, offered the bees homes in the forms of hollow clay tubes, baskets, clay pots or hollow logs. They found the bees readily accepted and occupied these highly desirable residences. For generations British beekeepers fashioned straw into skeps and used these as homes for their bees. In the summer a swarm would be caught and put into the skep, which became its home. The bees constructed the comb which they used for breeding and the storage of honey and pollen. In the autumn the beekeeper selected some colonies which he retained for breeding the next year and plundered the honey

from the remainder, killing the bees in the process.

Being rigidly fixed there was no way the beekeeper could inspect each comb in his skep without destroying the whole nest. Then, in 1852 the American Lorenzo Langstroth achieved what was undoubtedly the first real breakthrough in beekeeping when he invented and patented a movable comb hive. His idea was to encourage the bees to build their comb inside wooden frames, which were made so that when they were fitted into a hive the slabs of comb formed a nest similar to naturally drawn comb. The big advantage of having movable comb frames is that the beekeeper can remove any frame when required to check on breeding patterns and for the presence of stores or disease. Other hives had been invented before with parts which could be removed for these purposes, but none had been very successful. The bees usually either stuck the parts together with propolis (a resinous substance from sticky plants used by bees as a type of glue) or built additional comb in any spare space, making removal difficult. Langstroth realised that bees in natural conditions leave a space of between six and nine millimetres between adjacent combs. The dimensions of this space, usually called the 'bee space', are of critical importance. If the gap is less than six millimetres the bees will propolise the surfaces together. A bee space greater than nine millimetres will be filled by the bees with additional comb, usually called 'brace comb'. Langstroth designed his frames so that when they hung in the hive there was the correct bee space between each honeycomb and its neighbour, and between the frame ends and the inside walls of the hive, apart, that is, from suspension points of the frames. The bees respected this space and, with the Langstroth hive, frames could be removed with comparative ease.

A few years after Langstroth's hive appeared the idea of 'foundation' was conceived and developed. Foundation is the name given to thin sheets of beeswax which have a very accurate impression of the base of the cells pressed into the wax during manufacture. When placed into an empty frame and put in the hive the bees are encouraged to 'draw out' the foundation. To do this they turn it into comb by producing more wax themselves and using it to

Making a straw skep

extend the cells on both faces of the frame; the foundation is used as a central midrib. Offering foundation in this way speeds up the drawing out process and encourages the bees to construct nice straight comb. Most sheets of foundation used today have thin wires embedded in them to support the weight of the comb when it is full of honey.

Types of hive — which should you choose?

All modern beehives make use of Langstroth's idea of movable frames and the principle of the bee space. The Langstroth hive developed from the original has become the most popular hive in the world and in some countries, notably the United States of America, it is used almost exclusively. We in Britain like to do things our own way and much less standardisation has taken place, so that at present British beekeepers are faced with a wide choice of hive types. As well as the Langstroth we have the British National, Modified Commercial, Smith, Modified Dadant, WBC and, less commonly, hives such as the Catenary, Cottager and Jumbo are also in use. Each of these hives needs the correct size of frames to fit inside it and the frames from one will not normally fit in another hive. The exceptions are the British National, sometimes called Modified National, and the WBC. Both of these use a British Standard frame. This is probably the reason these hives are so popular in Britain. Although very different in appearance, most British beekeepers use one, or both of them.

Hive design is a very emotive subject, with many beekeepers arguing persuasively for their own choice of hive. Different hives are popular in different areas, for example, the Smith is used more in Scotland than in England, and the Langstroth is locally popular in some areas. Personally, I regret the lack of a standard hive, and was quickly persuaded to use the most popular type in my area. I suggest you do the same. There is much to be said for using the same as your beekeeping friends' and neighbours'. You can borrow odd bits and pieces of equipment, purchase second-hand equipment more often and sell your own surplus equipment more easily.

Hives in use today can be divided into single-walled and double-walled types. There have been several double-walled hives in use in the past, but the WBC is now the only one easily obtainable. All the other hives are single-walled types and there are more similarities than differences since they are all made up of the same parts; the differences lie mainly in the various dimensions.

The National and Langstroth hives illustrated on pages 16 and 17 both have a floor incorporating an entrance for the bees. The entrance can be partially or completely blocked by altering the position of a separate entrance block. Above the floor is a box usually called the 'brood body' or 'brood box'. In this box are hung frames in which the bees build comb. The queen lives in this area and this is where all the young bees are reared. Covering the top of the box is a queen excluder of either slotted metal or accurately spaced wires. The queen, being larger than the workers, cannot pass through the excluder and thus can only lay eggs in the brood body, whereas the workers can pass through and store honey in the frames in the boxes above. These upper boxes are called 'supers' and they can be added to the hive as required for honey storage.

During a good nectar flow a hive might need three or even four supers. On the top of the top box is placed a crown board, the purpose of which is to limit the upward heat losses from the hive. The crown board must have some holes in it to permit ventilation of the hive and in most cases standard sized slots are cut to allow Porter bee escapes to be fitted (see p. 48). The hive is topped by a roof usually covered with a water-proof material, such as thin sheet metal or roofing felt. The roof must have ventilation holes that are protected, usually by metal gauze, to stop bees and wasps entering the hive through them.

In the WBC the brood body and supers are made of thinner wood and another thin wooden outer wall surrounds them.

The best hives are made of western red cedar which lasts for years, even without treatment. This timber is, however, expensive and cheaper hives made from deal can be obtained. These are heavier than cedar hives and the timber needs treating regularly with wood preservative, but they are quite satisfactory.

Hives can be purchased ready-made, but many people prefer them 'in the flat' as they are cheaper in this form. The pieces arrive in a box with instructions for assembly and a supply of nails. The wood is, or should be, accurately machined to size and the hive should be relatively easy to make up, even for first-time hive builders.

DIY hive-making is relatively simple for those with some woodworking skills. Plans for the construction of certain hives can be purchased from the British Beekeepers' Association (the free MAFF leaflets on hive types give drawings and full dimensions), but possibly the best way for the do-it-yourselfer is to borrow a hive, a piece at a time, and copy it! (See p. 108.)

The shapes, sizes and dimensions of the various hives and frames are given in the Appendix (p. 107)

I advise the reader to write to a hive manufacturer for his catalogue where full details of this sort and a wealth of other detailed information are given.

The WBC

This hive, named after its inventor William Broughton Carr, is the only double-walled hive in common use today. The frames are housed in boxes, one brood box where breeding takes place and up to three supers on top, and these are surrounded by telescopic outer wooden walls called 'lifts'. The lifts, brood body and supers are supported by a floor and covered by a roof. This type of hive, often painted

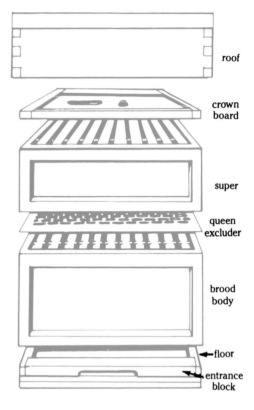

National hive

white, is what the general public understands by a beehive. Because of their double walls WBC hives are quite expensive when bought new, are said to be difficult to operate, and are certainly difficult to move with bees inside. On the other hand, many are still serviceable after years of use and can sometimes be purchased second-hand very cheaply. The WBC brood body takes only ten British Standard frames, which is probably insufficient breeding area for a prolific Italian-type bee, but for the average beekeeper with average bees I am sure that the WBC must be worthy of consideration, especially by those who prefer a 'proper beehive' to a 'packing case'.

The National

This hive is, without doubt, the most popular hive in Britain today. The brood body takes eleven British

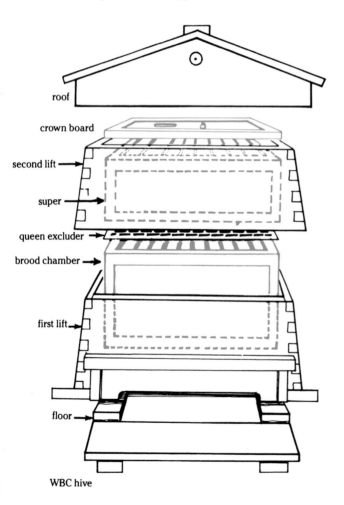

WBC hive

Standard frames, which appear adequate for most British bees, and the rebated walls on two sides give useful hand holds when moving the hive, which is a major consideration for beekeepers who follow the crop by moving their bees to oil seed rape, beans or heather.

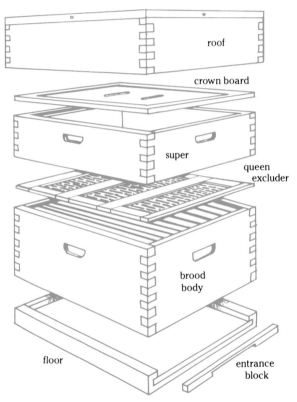

Langstroth hive

The Langstroth
This is the world's most popular hive, but has not caught on with British beekeepers — it is better suited to the more prolific Italian type of bee. It can be purchased with extra deep 'jumbo' brood body which should make it large enough for any bees likely to be found in Britain.

The Smith
This was designed as an economical hive and is much used in Scotland. The brood body takes eleven British Standard frames with short lugs, and because of its smallness it is probably the simplest to move with bees inside.

The Modified Commercial
The dimensions of this hive are similar to those of the National, but because it has straight sides and a deeper brood body eleven, and sometimes twelve, larger frames can be fitted, thus giving more breeding area for the queen. The frames are different from the British Standard but many parts are interchangeable with the National. It is suitable for the more prolific bees.

The Modified Dadant
This hive is the largest used in Britain. It is similar to the Langstroth, takes eleven large frames, is suitable for the most prolific bees but is not good for the beekeeper with a bad back.

Frames

Each type of hive requires frames of the correct size to fit. Brood frames, to fit in the brood body, are deeper than super frames which fill the super. All frames have the correct bee space between their sides and the inside walls of the hive, but frames have to be spaced so as to leave a bee space between adjacent comb faces. This can be accomplished in a variety of ways. The older method of spacing the British Standard frames was to add a metal frame spacer to the lugs of each frame. These frame spacers, called 'metal ends', are also made in plastic now and are often given the unlikely name of 'plastic metal ends', as well as the correct name of 'plastic end spacers'. A more modern idea is the Hoffman-type self-spacing frame. In this, the side bars are widened at the top to give the correct spacing and one side is machined to a sharp edge. When in the hive this arrangement means that there is only a small surface for the bees to stick together with propolis. Hoffman frames are standard for Langstroth, Modified Dadant and Modified Commercial hives and optional for the National, Smith and WBC. Many beekeepers who have kept bees happily for years using metal ends have never been tempted to convert to using Hoffman frames.

Manley super frame and Hoffman brood frame

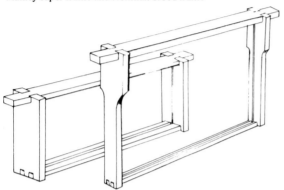

Making up a frame

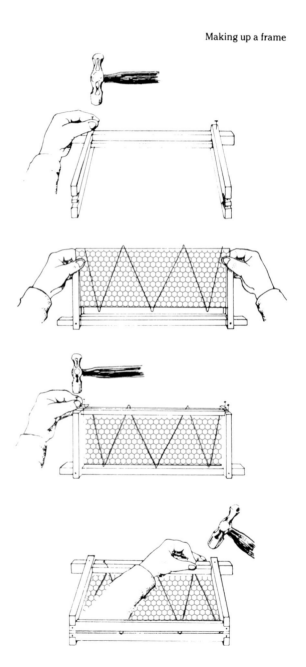

You, on the other hand, are strongly recommended to use nothing else in the brood body. Buy the Hoffman type when buying new and convert any existing frames to be compatible with them. This can be done by adding a plastic 'Hoffman converter' to the side bar of each ordinary brood frame or by adding two Yorkshire spacers to each. Either operation can be carried out with bees still on the frame, but they do object a bit!

Frames in the super are usually spaced wider apart so that fewer frames are required to fill the box; wider metal ends and plastic end spacers are available. Another type of super frame, called the Manley, has rectangular side bars and is very popular. Bees are less inclined to propolise frames in the supers, so the Manley frames are quite satisfactory.

Another system of frame separation used in both brood body and supers is the 'castellated frame spacer'. With this system the frames hang in accurately spaced slots in metal strips fixed to the ends of the hive.

All this must be most bewildering to a beginner. However, to a large extent your choice of frame is decided by choice of hive. If your chosen hive takes only Hoffman self spacers you are in luck. If you chose the Smith, WBC, or National, try to change to Hoffman type as soon as possible for the brood body and Manley type for the supers. You will not regret it.

It is strongly recommended that you never buy second-hand frames since these can carry disease.

Left: top bee space; **right:** bottom bee space

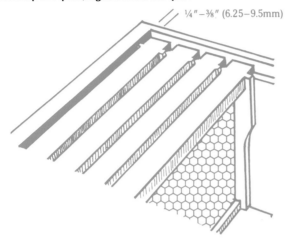

¼″–⅜″ (6.25–9.5mm)

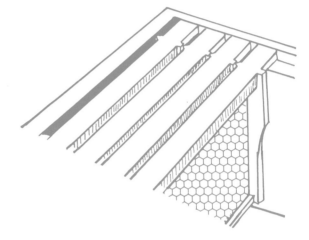

Having said that, most of us do just this when we buy our first bees, but it is good practice to stick to new frames and foundation thereafter.

Top bee space and bottom bee space

As well as leaving a bee space around the ends and the sides of the frames and combs, the design of each type of hive must ensure that there is a single bee space between the top of the frames in one box and the bottom of the frames in the box above. If more than a single bee space is left the bees will build brace comb in the gap, the boxes will be difficult to separate and the bees will take exception to your efforts to do so.

Some hives, notably the Langstroth, Smith and Modified Dadant, are built with top bee space. This arrangement means that the tops of the frames are one bee space beneath the tops of the boxes, so that if a flat board were placed on top of the brood box the bees would have room to walk about beneath it.

On the other hand, if the box were placed on the flat board, the bottoms of the frames would just touch the board and no bees could pass underneath. Other types of hive have a bottom bee space; the tops of the frames are level with the tops of the boxes and the bottoms of the frames are one bee space above the bottom of the box. In this case, a box standing on a flat board would enable the bees to walk around on the board below the frames. I don't think the bees mind much one way or another, but users of hives with top bee space claim many advantages for their system. The main ones are that the queen excluder is propolised less by the bees, and that one box can be slid on top of another without decapitating bees on top of the frames. The disadvantage is that the queen excluder on a top bee space hive must be framed and of rigid construction to prevent it sagging. These types are more expensive than the slotted zinc sheet types which rest on the top frames of the top bee space brood body.

Protective clothing and equipment

On the subject of protective clothing the beginner is given a good deal of conflicting advice. Mine is to cover yourself from head to foot and let the bees do their damnest. I see no virtue in getting stung unnecessarily. If, after a time, you discover bees do not bother you, then shed clothing and meet the bees face to face, but let this be your decision not the bees' or anyone else's.

The first item you require is some type of *bee suit*. Specially made ones can be purchased, or boiler suits can be modified. Zips are better than rows of buttons and velcro strips on sleeves and trouser bottoms give extra protection. Most lady beekeepers take to trousers when handling bees, as nylon tights offer little protection against the sting of a determined honey bee. Bees are said to dislike clothing made of thick wool or anything blue in colour. White or olive green, on the other hand, seem to keep the bees happy and beekeeping clothing is often made in these colours. White has the additional advantage of reflecting heat and keeping the wearer cool.

Next come wellington *boots*. These form an effective barrier to attack at ankle height (a favourite point of attack). Rubber boots are hot in summer and leather boots which cover the ankles are more comfortable if they are available.

Gloves – these protect your hands against stings and prevent the propolis sticking to your fingers. Many beekeepers do not wear gloves, but I believe in being

prepared, and suggest you do the same. For some operations the beekeeper must remove his gloves, though again I see no virtue in a beginner getting stung unnecessarily. Old hands who 'never' wear gloves usually have a pair 'just in case'. They must also be prepared to clean off the propolis from their hands with methylated spirit or live with grubby hands, since propolis does not wash off easily. Rubber washing-up gloves are adequate, but special gloves are sold in either thin leather or plastochrome with gauntlet extensions which cover the cuffs and reach half way up the forearm.

The *veil* and *hat* – these come in many shapes and forms. All veils have a mesh of thin nylon, cotton or wire small enough to prevent the bees coming through. Some types incorporate a hat, some require a hat of some sort to keep them in position, and some are self-supporting and zip onto a special bee suit. Study the manufacturers' catalogues and choose to suit your pocket, or study a veil and hat carefully and make your own. The most stunning outfit I have seen, worn by a very attractive and fashion-conscious lady beekeeper, was home-made, with stiff nylon net reaching up to an elegant and elderly Panama hat.

Here again some beekeepers say they 'never wear a veil'. I suggest you wear one always. A sting on the forehead or temple can close both eyes for a day. It is amazing how amusing other people find this, except those you work with, who sometimes regard it as a self-inflicted injury designed to inconvenience them in some way.

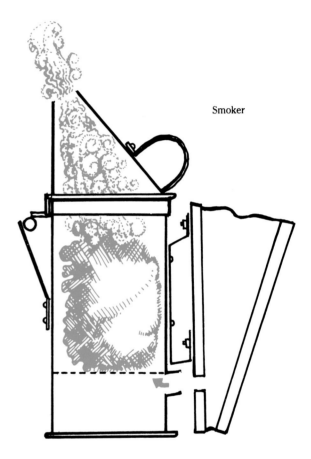

Smoker

Hive tool

Hive tool – you need something to prize apart the boxes and frames of the hive, which will become lightly stuck together with propolis by the bees. A wide-bladed screwdriver or chisel will suffice and with care will not damage the hive parts too much. A specially designed hive tool is better, of the type sold by all equipment suppliers.

The smoker – smoke, in moderation, subdues bees and so a smoker is essential. They are not easy to make and not cheap to buy. The smallest and cheapest are the straight-nosed variety, which are difficult to light and easy to 'let go out'. I strongly recommend the bent-nosed smoker: it is larger, will hold more fuel, and stays alight with less attention and frustration. For fuel smokers, burn anything from corrugated paper or dried grass to old rags and hessian. Some fuels burn quickly, others more slowly. Decayed and dried wood is very good. It takes some lighting but will smoulder for ages and produces a cool smoke which the bees seem to prefer. Lighting other fuels is easier, e.g. roll up your hessian, or whatever you are using, to a size which will fit inside

Lighting the smoker using newspaper and dry grass

comfortably. Light some newspaper and pop inside the smoker, puff to get a good flame and slowly lower the fuel into the cylinder, puffing strongly. Close the lid when the fuel is burning well and keep puffing from time to time. Practise once or twice away from the bees to get the hang of the thing. When you have finished put a cork in the spout and leave outside for the fuel to stop smouldering and the smoker to cool down.

Other equipment will be mentioned later, which is either desirable or designed for special purposes, but a beekeeper equipped and clothed with the items above can expect to cope with most situations he is likely to meet in Britain, and come out on top.

Left to right: queen, worker, drone

Workers

Worker bees make up the vast majority of the bees in the hive. They are present throughout the year, although their numbers wax and wane. A colony with, say, twenty thousand bees in January can steadily expand to about fifty thousand in June or July and decline to twenty thousand again by Christmas. Workers are incomplete females in that they have lost the ability to lay eggs. Instead, they have developed other specialised functions. Depending on age, workers clean up and remove debris from the hive, feed the brood (the young bees), feed and groom the queen, produce wax, build comb, collect and store nectar, pollen and propolis, guard the hive, fetch and distribute water and cool the hive.

Workers start life as a fertilised egg laid in a worker cell; three days later the egg hatches into a grub or larva. The larvae are fed 'bee milk', or royal jelly, for two days. This highly nutricious milky-white substance is produced by the workers' hypopharyngeal glands. After two days the larvae are weaned progressively onto a diet of nectar and pollen until by the eighth day they have grown to fill the bottom of the cell. On the eighth or ninth day the workers seal over the top of the cell. Inside, the larva spins a cocoon around itself, spends the next thirteen days changing completely from a grub to a pupa, and then, on the twenty-first day after the egg was laid, it chews its way out as an adult bee.

At this stage the bee is wet, hairy and light grey, and cannot sting or fly. She will spend the next three weeks inside the hive before venturing forth into the outside world. Firstly, she will clean and polish her own cell ready for the queen to lay another egg and then, when about three days old, she will feed the older brood on pollen and nectar and when her own hypopharyngeal glands develop she will feed the younger brood. When about twelve days old her wax glands develop, she produces wax and builds comb.

At about eighteen days the sting has developed, and she guards the entrance to the hive until at about twenty-one days she makes her first trial flight. From then on she will, weather permitting, be out foraging for nectar, pollen, water or propolis. In summer she can literally work herself to death in three weeks from her first flight. On the other hand, bees hatched out in the autumn go through the various stages of development but many of their tasks are not necessary at that time. Their development is, therefore, suspended and most survive the winter to commence the flying stage of their lives when spring arrives.

The queen

There is usually only one queen bee in a colony. She is much longer than the workers – she has longer legs and a much longer abdomen. She has a sting without a barb which she will use on other queen bees, but rarely on the beekeeper. She is the only fertile female in the colony and her main function is to lay the eggs which will hatch out into young bees, leaving the care of her offspring completely to the workers. She usually starts life herself as a fertilised egg laid by the queen in a specially constructed queen cell. The egg hatches into a grub and is fed copious quantities of royal jelly for the whole of the five days of the larval stage. The queen cell is sealed by the bees on day eight, the queen pupates and emerges fifteen days after the egg was laid. If the new virgin queen is to head the colony she will, in a few days, learn to fly and, usually within a week of emerging, she will go out on a mating flight. She will mate, probably with several drones, on the wing away from the hive and if all went well she will return to the hive and start to lay within a few days. She will only rarely leave the hive after this and appears to prefer a dark place to hide away. She can live for several years but will be replaced by the workers as soon as they decide she has passed her prime.

The queen's rate of lay will depend on the seasons, but at the height of the breeding season she can lay over two thousand eggs a day. She can also lay two

Guard bees and foragers

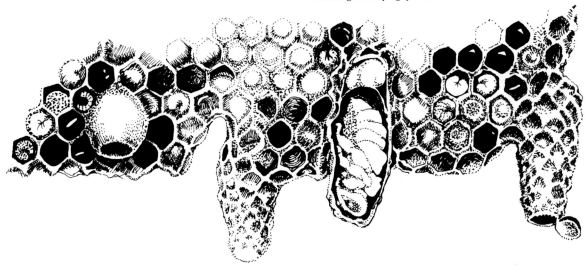

Left to right: queen cup, sealed queen cell, section through cell showing developing queen, used cell

different sorts of eggs, fertilised eggs which usually end up as workers, but can end up as queens, and unfertilised eggs which end up as drones. The drone eggs are laid in drone cells which are slightly larger than worker cells and it is thought that as she checks the cell she measures the size with her feelers and decides which type she should lay. She lowers the rear of her abdomen into the cell and the egg starts its journey from her ovaries. The sperms from the males with which the queen mated are stored in the spermatheca for the whole of the life of the queen and as the egg passes the entrance of the spermatheca it either receives a sperm and is laid as a fertilised egg or passes straight by and is laid unfertilised.

She has a second function which is to produce on the surface of her body a special 'queen substance'. The workers lick this off her in the grooming process, and in the normal food sharing the substance is passed around from bee to bee. The quantities which each bee receives are minute but of great importance in maintaining the cohesion of the colony in that it appears to inhibit the impulse to swarm.

Even though she is considerably larger than the workers the queen can be surprisingly difficult to spot amongst some thirty thousand other bees. For that reason many beekeepers, having once found her, put a dot of quick drying paint on her thorax to enable her to be found more easily on later occasions (see p. 26).

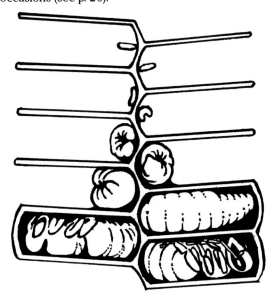

Section through cells showing developing bee

Marking the queen

From the start of May to the end of July you will need to inspect your bees regularly to prevent them from swarming. At some stage you will probably need to find the queen. This can be difficult at any time, but at the height of the swarming season it can often be a frustrating, time-consuming and thankless task. It is far simpler to find her in early spring when there are fewer workers for her to mingle with, and having found her, to mark her and so make her easier to find on later occasions. To do this she is caught and a single drop of quick-drying paint is applied to the top of her thorax. The paint is allowed to dry for a minute or two, and she is then released back into the colony. By changing the colour annually it is possible to 'date stamp' your queens. The internationally agreed code of colours for queen marking is as follows:

For years ending in:	Colour:
1 or 6	White
2 or 7	Yellow
3 or 8	Red
4 or 9	Green
5 or 0	Blue

For example, a queen raised in 1979 would be marked with green paint, and one in 1983, red.

There is, of course, no need at all for you to follow this convention if you do not feel the need to buy five separate pots of paint. Marking all your queens white is a good idea if you have any trouble with your eyesight, since it is far easier to pick out a queen marked with white than with any other colour. Special queen-marking paint kits can be purchased, but a miniature tin or jar of the kind of enamel used by model makers works well. A matchstick makes a good applicator. The end holds sufficient paint and makes a dot of the right size. The queen, not unnaturally, appears reluctant to stand still for this operation and various queen marking aids can be purchased to make the job easier. Probably the simplest to use is the 'press in' queen cage. When the queen is found she is covered by the cage and manipulated onto an area of the frame covered by stores. Then the cage is pressed downwards, the spikes go straight into the honey comb and the thin threads hold the queen against the comb surface. The thorax can easily be touched with the matchstick, the paint is allowed a minute or so to dry and the cage is then removed.

More experienced beekeepers simply pick up the queen, holding her by her thorax between finger and thumb while they mark her. This is not a recommended course of action for beginners since queens can easily be dropped or injured and an injured queen early in the season can be a disaster. I found this out myself the hard way and my friends have never ceased to remind me of the time I gave one of our queens a partial hysterectomy. She was not amused — she didn't live long either.

If there is any chance that you will need to mark a queen during an inspection make sure you have all the necessary equipment to hand before you open up the hive. There is nothing so frustrating as having to break off to look for a pot of paint when you have finally run her to earth.

A useful piece of advice is to practise on drones first before tackling the queen. Drones are expendable and of much less importance than the queen. If you do this, however, you must be prepared to destroy the drones afterwards or you might be faced with a frame bursting with 'queens' on your next inspection.

Drones

The drone we all know is an idle fellow who does no work at all! The only male in the hive, the drone starts his life as an unfertilised egg in a drone cell. After three days the egg hatches and the larva is first fed 'bee milk', and then later pollen and nectar for a longer period of seven days. At ten days the drone larva is sealed in the cell. The drone cell is larger than the worker cell and is sealed over with a convex domed capping. The larva pupates in the cell and the fully developed drone emerges at twenty-four days from the time the egg was laid. The drone is looked after completely by the workers, even being fed by them, and does no work at all. His sole function is to mate with a virgin queen should he have the opportunity to do so. If he does fulfil this role he dies shortly afterwards.

Marking the queen using a queen cage

No drones are needed when virgin queens are not being reared, so to save food the drones are not allowed to live through the winter. At some point in late autumn the drones in the hive will be starved of food and driven out of the hive where they perish — it happens surprisingly quickly.

In early spring the queen will lay one or two drone eggs daily and then a few more each day as the season progresses. During the summer months there may be a few hundred or a few thousand drones in the hive. Very few of these will ever fulfil their sole aim in life but the workers seem to like to have many more than necessary in reserve, and will pull down worker comb and rebuilt it to breed as many drones as they require.

The drone is bigger than the worker, is squarer in the tail area but has no sting, and has a larger set of wings which cause him to make more noise in flight. Drones from one hive are welcomed into other hives and during the mating season they congregate together on the wing into drone assembly areas which are the starting points for the race of the kamikaze mating game when a virgin queen turns up.

Senses and communication

We know from the writings of William Shakespeare and others before him that for generations man has been fascinated by the organisation and efficiency of the honey bee colony. A close study of the bees going about their tasks elicits a profound admiration for the system of organisation whereby each and every bee appears to work diligently, yet unsupervised, at her appointed task, with no squabbling over choice of jobs, fighting over food or territorial disputes. The colony exists for the benefit of each bee: each bee is as important as her sister, food is collectively gathered, stored and shared out and never denied to a hungry bee while any remains.

All these qualities which we humans find so commendable have evolved in the honey bee colony over the ages. Colonies which were least successful at gathering, storing and distributing food would perish in the winter and their places would be taken by their more successful and adaptable relatives. Over the eons certain characteristics, behavioural patterns and instinctive reactions have become fixed so that the pattern of behaviour for each adult worker is decided and fixed from the moment she chews her way out of her cell.

The bee is well provided with sensors for information gathering. As well as two large compound eyes, the bee has three other simple eyes. Together, they enable the bee to see and recognise definite shapes, and they are sensitive to polarised light. Bees cannot see red as we know it, but are able to see light in the ultra-violet region of the colour spectrum which human eyes cannot detect.

As well as providing information on touch, the twin antennae also house smell and taste sensors. Bees are deaf, yet sensitive to vibrations, so can sometimes respond to loud noises which may cause parts of the hive to vibrate.

Worker bees also have another specialised mechanism for the distribution of information, called the 'Nasonov gland'. A row of workers can often be seen in the hive entrance, facing inwards with heads down and bottoms up, standing still with their wings fanning at a terrific rate. The last segment of the abdomen is dropped and the Nasonov scent gland is exposed. The scent produced is driven backwards by the fanning and acts as a sort of homing aid to other bees, which detect and identify the scent and follow it down to reach the hive.

Inside the hive the allocation of tasks is decided mainly by the age of the bee and the stages of its normal development. For example, a young worker bee will feed the younger brood while her glands are actively producing brood food and will move on to cell building when her wax secreting glands come into action. The natural progression of tasks with age is from cell cleaning to older brood feeding, then younger brood feeding, through comb building,

Bee anatomy
a Hypopharyngeal gland   b Oesophagus   c Honey stomach
d Ventriculus   e Malpighian tubules   f Rectum   g Wax glands
h Nasonov scent gland   i Venom sac   j Sting

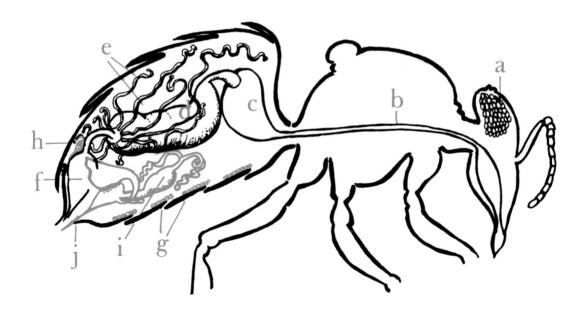

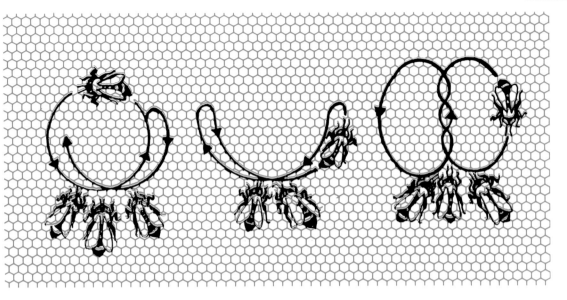

The bee dance
Left to right: the 'round' dance, food within 10 metres; the 'sickle' dance, food further than 10 metres; the 'wag-tail' dance, food further than 15 metres.

entrance guarding and finally to foraging duties. This pattern does not, however, allow for any abnormal circumstances or variations, for example, when for one reason or another no brood is being reared or when a great increase in comb building is suddenly required, e.g. after swarming.

From observations of marked bees it would appear that each bee spends a great deal of time just walking around, collecting information with her sensors. In this way the bee will apparently find out what wants doing and will promptly start to do it. Thus the bees are able to detect very quickly sudden changes in the environment, and they react accordingly. A study of marked bees also brings to light the surprising fact that bees spend quite a lot of their time doing nothing at all, just resting. Not exactly the image of the busy bee, but the resting bee is using up little food, providing heat to maintain the brood-rearing temperature and is available in reserve to react quickly to changing circumstances.

Possibly the most amazing feat of communication is that used by a worker bee inside the hive to indicate to other bees the distance and direction of nectar sources outside. Professor Karl von Frisch is credited with the discovery that bees apparently doing some sort of a square dance on the surface of the comb are, in fact, passing information to other bees. There are two different and distinctive bee dances: the round dance and the wag-tail dance.

If a worker finds a good source of nectar close to the hive, within 80 to 100 metres, she will, on returning to the hive, immediately perform the round dance, running round in a small circle first to the left then to the right and repeating the dance until she attracts sufficient recruits to work the new

Bee fanning

food source. Other workers, aware of the taste and smell of the nectar, leave the hive and search around nearby until they locate it.

When the nectar source is further afield, however, more accurate information on its whereabouts must be given or the bees would spend wasted time and effort searching for it. In this case, the information is passed in the wag-tail dance. The dancing bee runs in a straight line on the surface of the comb for a small distance, wagging her tail vigorously from side to side. Then she circles to the left to return to the starting point, does another straight tail-wagging run, circles to the right and repeats the whole cycle. These actions supply very precise information to other bees in a truly amazing way. If the central run of the dance is straight up the comb, from bottom to top, it means that bees need to leave the hive and fly directly into the sun to find the nectar. If the dance is reversed, with the straight part going down, from top to bottom of the comb, the bees need to fly from the hive directly away from the sun. If the straight run is thirty degrees to the right of the vertical, going towards the top bar, the bees leave the hive, locate the sun and fly at thirty degrees to the right of it to find the food.

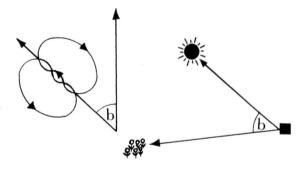

Angle 'b' between the 'wag' and the vertical of the comb tells the other bees the angle between the sun and the nectar source

Potential foragers know what food to look for from the smell and pollen carried by the dancers, and information on the distance of the food from the hive is given by the pattern and intensity of the dance. The more dances completed in a set time, the less is the flying time necessary to reach the food source; the longer the dance takes to complete, the further away is the food and the longer the flying time to reach it.

It is now believed that these two dances enclose a whole series of intermediate dances. The round dance proper is probably only used to indicate a food source ten metres or less from the hive; this is followed by a type of transition dance, sometimes called a 'sickle' dance, used to indicate food between ten and fifteen metres away. When food is over fifteen metres traces of tail-wagging appear and increase steadily until at a hundred metres the 'round' dance movements have almost disappeared.

This wonderful example of communication is, however, only part of an operation of mind-bending complexity. The sun is moving all the time, so information given relating to its position soon becomes out of date. The dancing bee allows for this and predicts its future position when she passes on her information. The bees' ability to see polarised light is utilised to permit information about the sun to be passed on even in dull and cloudy weather when we humans haven't seen the sun all day. Moreover, a foraging bee setting out to fly in a given direction must allow for the wind which could blow her off course. For example, if she needs to fly thirty degrees to the left of the sun and the wind is blowing from her right, she will head in the chosen direction and then fly pointing just slightly to the right to allow the wind to blow her back on her chosen flight path.

No short summary such as this can do justice to the bees' ability to communicate. The more one reads, the more one is impressed. Perhaps the most amazing feat of communication occurs after swarming when foraging bees, whose experience has been limited to finding and collecting nectar, pollen or water, suddenly become surveyors. They inspect every nook and cranny to assess its potential as a future home for the swarm, and then report back their findings by performing the tail-wagging dance on the uneven surface of the hung-up swarm! Reports from several bees are collected and somehow compared, and then the swarm will break and the bees will fly directly to the most suitable home available.

The sting — the bee's only defence

If you want to keep bees merely for the joy of watching them go about their business of gathering nectar from the surrounding countryside, you may go for years without ever finding out at first hand about 'the sting in the tail'. If, however, you intend, as most of us do, to rob the bees of the fruits of their labours then you will quickly find that the bees have a very effective means of defending their property. Having said that, a bee's sting is usually nothing to worry about. Most people having been stung, feel pain only for a minute or two and experience a swelling of the surrounding area. The swelling usually looks worse than the actual hurt itself, but in some cases the embarrassment is as bad as the pain. I am thinking particularly of a sting in the area of the eyes, which turns even the most attractive female into a Miss Piggy in no time at all!

Without a sting bees would have become extinct thousands of years ago. Any large collection of a sweet commodity like honey is likely to come to the attention of predators, be they animal, insect or bird. Without any means to protect their stores, the bees would lose everything and starve during the winter months. The bee does not use its sting against humans lightly, as she cannot withdraw the sting and in her efforts to do so will pull the entire stinging mechanism from her body and will die shortly afterwards. Bearing this in mind, there is normally a natural reluctance for bees to sting at all. Experienced beekeepers who know their bees and understand the effects of the weather and the normal hive conditions dispense with veils and seem to lead a charmed life with immunity from stings. You, as a beginner, are not advised to emulate them, since you are much more likely to cause the bees to sting and are more likely to react when stung. Only the masochists among us will revel in being stung unnecessarily, and with proper protection there is no reason why a sting should be other than a relatively rare occurrence.

The worker bee's stinging mechanism is hidden in a cavity in the rear section of the abdomen. When attacking, the bee arches her body and out shoots a slightly curved, barbed, tapered and sharply pointed tube which is thrust into the enemy. Inside the cavity and connected to the tube are two large muscles which operate like a pump to force the poison from a venom sac into the victim. After stinging, the bee usually tries to withdraw the sting but is usually unable to do this from human flesh. In her frantic efforts she tears out the stinging mechanism from her abdomen and is so badly injured that she will soon die. The muscle mechanism continues to pump in poison even after this has happened, so obviously the victim should remove the sting as soon as possible. The best way to do this is to scrape the sting out by pushing the sharp edge of the hive tool along the skin with a chiselling action. The thing not to do is to try to remove the sting by grasping it and pulling, since this forces in more poison. The normal human reaction is to rid oneself of the sting as soon as possible and in the case of stings this is the best possible initial treatment.

The effect of a bee sting on the upper lip

Scratching out a sting

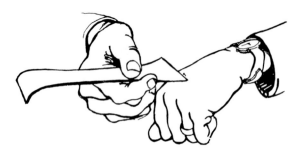

The chemistry of stings and the medical aspects of human reaction to stings are an extremely complex matter. My own probably over-simplified understanding is as follows. Bee venom is a mixture of protein and other substances; it is mainly acidic, but alkaline substances are also present. When the sting enters the body two natural antibodies are produced to neutralise and remove the venom. If your body tends to react to stings by producing more of the antibody Immunoglobin G than Immunoglobin E you should have no problems. This antibody copes with the venom, has no nasty side effects, and over a period of time the antibody level in the body builds up so that very little reaction to bee stings is then noticed. Most people come into this category, certainly all the hardened old bee masters who scorn protective clothing of any sort and carry on handling their bees oblivious to any attack.

It is when your body produces more Immunoglobin E than G that problems arise. This antibody reacts to stings by producing histamine which limits the effects of the venom, but in so doing causes swelling in the area of the sting. In most people this is all that happens, but in a very small number of people it can also cause much more frightening effects, from giddiness to difficulty in breathing followed by unconsciousness. Obviously if you or any of your family suffer from such a serious allergy to bee stings you will take medical advice, but my advice is to give up beekeeping and look for another hobby. It just is not worth taking risks with one's health and happiness.

Another problem can occur when the balance between the two antibodies changes at a later stage during one's beekeeping career. It appears that in a few beekeepers who have a high Immunoglobin G level and who therefore appear immune to stings, the Immunoglobin E level slowly builds up over a period of time and at some point reaches a high level so that a single sting will cause most of the serious reactions previously mentioned.

Beekeepers who know they suffer from an allergy to bee stings often carry antihistamine tablets, such as Piriton, which they take immediately they are stung. Some creams, such as Anthisan, when rubbed on the affected area soothe the soreness and reduce the itching. Apart from this most other external cures for bee stings have little or no effect apart from the possible psychological advantage in taking some action to reduce the pain.

My advice is to try to avoid being stung wherever possible. Wear proper light-coloured protective clothing, gloves and ankle protection when handling your bees. Move slowly with no jerky movements. Do not go near your bees when thunder is heard since this tends to make them bad tempered. Choose the times for inspections carefully — when the bees are flying in reasonable numbers is a good rule. Do not keep the hive open too long in late evening. Bees are likely to be bad tempered when they are starving, when the smoker has no effect since they have no honey to feed upon. They also appear to be more bad tempered when they are queen-less, a condition which can happen from time to time, and they certainly are bad tempered when they have been taken to oil seed rape fields to work the flowers. Bee venom has a distinctive smell which seems to enrage other bees, so once stinging has started it is not easily stopped.

At the risk of boring you, but with no apology on my part since I think it is of vital importance to get it right, I will repeat my advice on coping with stings: avoid stings as far as possible; be prepared; wear correctly fitting bee-proof clothing, gloves and boots; choose the right times for inspections; use the smoker correctly, i.e. frequent small puffs not just one great choking cloud; do not knock or jar the hive or hive parts; move slowly and deliberately at all times. If you do not get over-excited, then neither will the bees.

Hives are kept and managed by beekeepers in all types of places, from castle grounds to garage roofs. In the 1940s and '50s the country beekeeper was considered by the town-based beekeeper to be very fortunate. The countryside was full of leafy lanes bordered by verges aglow with wild flowers, long-term pastures covered with wild white clover, and hawthorn hedges abounded. Then came the mechanisation of agriculture with the farmer increasingly employing chemical sprays to destroy weeds and pests, and tearing up hedges and rough areas to make larger fields easier for cultivation. These days the country beekeeper has more problems than the town dweller. Efficient weed-free corn fields provide no forage for bees, roadside verges are often sprayed to kill weeds growing there, while some crops, like oil seed rape, can provide nectar in large quantities but bring many problems of their own. The town dweller, on the other hand, has seen a steady urban spread, with most houses having their own garden and most houseowners filling their gardens with attractive flowers and shrubs.

Nearness to parks or large country houses can be an advantage. Both can be expected to have numbers of mature flowering trees — maple, lime or horse-chestnut — all of which secrete nectar *at times*. The last two words are important. There is nothing certain about it: secretion depends on soil, climate, position, and most important of all, the weather during flowering time. If the weather is dreadful the plants just will not secrete nectar, and in any case the bees will not fly out in search of it!

In most places in England a colony of bees in a garden will find enough nectar to live on and bring in some surplus for the beekeeper. In fact, it is said that most sites could support five or six hives. The bees fly out in all directions on foraging trips. If they find food within a few hundred yards they can quickly return to the hive, unload and start out again. They burn up nectar on the flights and obviously the shorter the journey from hive to food source the less food is consumed by the bee. Bees can fly up to three miles in search of food. At this distance the bees would probably use up for themselves most, if not all, of the nectar they had gathered. One would normally expect bees to operate within a range of a mile from the hive, so it is important to check before you establish your hive to see how many other beekeepers there are around you, and how many colonies they have. For example, if your next door neighbour has six colonies already, you, or he, would need to find a site some distance away or all the colonies would suffer. This is a common practice. There are many beekeepers who keep all their bees in 'out-apiaries'. An advantage of this is that there are fewer bees around your own house or buildings to take an interest when you extract honey or handle the supers. The disadvantage is that everything has to be transported between home and out-apiary, extra travelling time is involved, and you always forget something!

In most cases the beginner will look first at his own garden as a site for a hive. Often a corner least suitable for gardening is most suitable for bees.

Wherever you position the hive the bees will be very active in front of it for a distance of up to a metre, so the direction a hive faces is important. If you have a corner bounded on two sides by a wall, fence or hedge, stand the hive facing into it and a metre from the corner. The bees flying out will then have to climb up to overcome the obstacles and will set off above head height, and neither you or your neighbours will be bothered by head-on collisions with returning bees.

The WBC hive has its own stand to keep it above ground level; this enables air to circulate freely beneath and keeps the bees drier and warmer. Other hives are best placed clear of the ground in the same way, either on specially made hive stands or breeze blocks, or frames of some sort. This also makes for less stooping during inspections. I use discarded milk crates, but your local dairy may not be as obliging as mine.

The hive and stand should be on a solid base just slightly tilting forwards from the back of the hive towards the front, so that any water entering the hive runs straight out from the entrance.

Bees do not like being positioned under trees which constantly drip on them, or under high voltage electric cables. Neither do bees like horses; the feelings are mutual, so avoid riding stables. Also, children who constantly play in the flight path just in front of the hive entrance will be stung sooner or later. Do not point the hive entrance onto a garden path or into the neighbour's garden or you will have trouble. If you have children and a sand pit keep the pit covered as bees like to suck moisture from wet sand and children don't like their company. Ideally, hives should be protected from winter winds and in a position where they catch the early spring sunlight.

Your neighbours might not share your interest in bees so do not place hives in prominent positions. They cannot complain about your bees if they don't know about them! Hives placed where they attract attention often become targets for vandals playing push-and-run. The bees often win this game in summertime, but in the winter or pouring rain the results can be disastrous for the bees and heart-breaking for the beekeeper. On the other hand, complete hives or sometimes all the bees inside them have been stolen on many occasions. A nasty thought as the thief will be a beekeeper himself and we like to think of ourselves as a friendly, sociable and utterly honest bunch.

From all of this it can be seen that the perfect site just does not exist and the choice will of necessity be a compromise. If you have any doubts ask a beekeeping friend to look the proposed site over — it could save a lot of problems later on.

Moving the bees in

Having procured your chosen hive and decided on the site it only remains to move the bees in and then you are in business. Your bees might come in any of several different ways.

A swarm
If you have arranged to take delivery of a swarm, your bees might arrive in anything from a straw skep to a cardboard box. More often the swarm will be pointed out to you and you have to collect it yourself. This can be simple, difficult or impossible, and you are advised to seek the aid of a more experienced beekeeper for this encounter. (See section on swarms and swarm collection, p. 79.)

In the case of a swarm an empty hive is brought in, the roof and crown board are removed, and the bees in the swarm are dumped into the brood body containing frames of drawn comb or, better still, foundation. An empty super placed on top of the brood body helps to stop the bees spilling out over the sides during this operation. Hiving a swarm is best done in the early evening. Replace the crown board and roof and the bees should go to work, produce wax, make comb and in a few days the queen should start to lay and you can call yourself a beekeeper — assuming, that is, that the bees do not swarm out again the day after you put them in. It is very good practice to feed the bees a bucketful of warm, thin syrup twenty-four hours after they are hived. This assists comb building tremendously. (See section on feeding, p. 58.)

A colony of bees (bees in the hive)
If you buy a full colony of bees, try to persuade the seller to help you move them. WBC hives are

Hiving a swarm the quick way using an empty super to guide the bees in

especially difficult to move when occupied. The roof and lifts are removed and the brood body and supers are lashed securely to the floor. Holes on the crown board are covered with a flat piece of board, and a thin piece of foam rubber is used to bung up the entrance. Great care is necessary as the hives are awkward to move, leak bees like a sieve, and the inexperienced helper who lifts the hive by the legs often finds they come off in his hand. It is a two-man job, and both should be veiled and gloved.

Single-walled hives are much easier to move. The entrance is blocked with a piece of foam rubber and the hive is first solidly fixed together and then lifted as a whole, roof as well. Various devices are available to fix hive parts together: staples and metal triangles are the simplest; cheapest, and worst, is a lash-up of

Lock slides, travelling screen and foam rubber entrance block

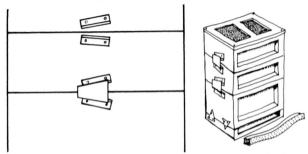

clothes line or baler twine. I personally favour a lock slide device sold by Steele and Brodie. They are not cheap, but are relatively safe and easy to fix and unfix.

The time to move bees is late evening when they have finished flying for the day. Moving bees at this time means that the closed-up hive does not get too hot and cause the bees discomfort and distress. If the distance to be moved is such that the bees will be closed up for more than an hour the crown board should be replaced with a small mesh-covered travelling screen, which prevents the bees coming out but ventilates the top of the hive. In extreme cases water can be sprayed onto the mesh of the travelling screen to cool the bees further.

A nucleus
A nucleus is a small colony (nuc' for short). It usually consists of five or six frames of drawn comb liberally covered with bees, including a laying queen. There should be brood of all ages (eggs, larvae and sealed brood), some pollen, and one frame should be fairly well stocked with stores.

Nuclei can be ordered from beekeeping equipment suppliers and will usually be delivered in May or June. In this case, the nucleus will probably arrive in a thin plywood, ventilated travelling box with full instructions on how to transfer and establish the bees in their new home. Be careful to order the nucleus on the right sized frames to fit your hive.

You might purchase the nucleus from a local beekeeper, in which case a good plan is to take your hive with you, install the bees in his apiary and bring them home in the hive as already detailed.

The advantage of starting with a nucleus is that

there are fewer bees around when you perform your first beekeeping operations and their numbers grow with your increasing confidence and handling skill.

The dummy board
This most useful piece of equipment is one which is often needed when moving a nuc' into a full sized hive, or when installing a swarm. It consists of a flat hanging board made to the same dimensions as a frame. If you have insufficient frames of foundation or drawn comb to fill the hive you simply push those that you have to one side, drop in the dummy board and push it up tight against them. This effectively seals off the rest of the hive and enables the bees to control more easily the temperature of the occupied part. Dummy boards really are a job for the do-it-yourselfer. They can be made quite easily from plywood and a spare topbar, or even more easily by pinning cardboard onto one side of an old frame.

Three feet or three miles

Once the bees are used to flying to and from a hive in a particular position they will return to that site. If you move the hive, bees will return to the old site

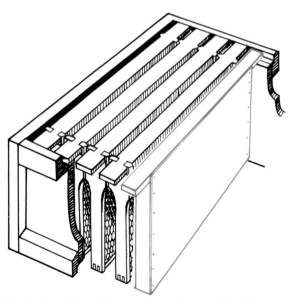

Section through brood box showing use of dummy board

and not to the hive. Therefore, if the bees you have just purchased were formerly kept within flying range (up to three miles), they will all return to the apiary of their former owner. If, on the other hand, the bees are moved more than three miles into new and unknown territory they will fly for some time round the hive, fix its position and then return there from then on. Bees lose the memory of their former site after a week or so, so if you wish to move bees you have purchased a distance of less than three miles you have to do it in two moves. They are first moved from the old site to another site over three miles away, left there for a week, and then moved from there back to your chosen site.

If you wish to move the hive around within the garden you can do it with a series of short daily moves. If you move it less than three feet (a metre) the bees return to where it was, fly around and locate it visually and fix the new location in a day or so.

All the above remarks relate to moving bees during the flying season. In the dead of winter bees can be moved around without any precautions, since they appear to relocate when they start flying in the spring.

Good luck in moving your bees. It can lead to some interesting encounters with onlookers, and remember the last check after moving: When you have re-assembled the hive always remove the foam rubber in the entrance. Imprisoned bees collect no honey — they just die.

This section is for those of you who obtained your bees during the summer months as a swarm or a nucleus. If your bees came as a full colony you should be starting weekly inspections for the purpose of swarm prevention (see section on swarming, p. 79). In the case of a swarm or a nucleus, you will need to open up the hive a week or so after installing them to check that all is well. In both cases you should have workers, drones and an active laying queen, initially occupying less than the full complement of frames and trying like mad to expand to fill the hive with bees before winter. Expansion means more brood reared, which gives more foragers to collect extra stores for the winter, and less chance of the colony starving, so the drive to expand is tremendous.

Choose a nice day for your inspection. You can look at bees any time when they are flying freely, but it is possible to chill the brood and retard expansion if you keep the frames out for too long in a strong wind, even on a nice day, so do not take your time about it. It should only take ten minutes or so at most to check either a nuc' or a swarm.

First light the smoker. Then don full protective clothing, find the hive tool and check to see if the smoker has gone out. Approach the hive and gently give two or three puffs of smoke into the entrance. Give the bees a minute to come to terms with the smoke, then gently give one more puff before removing the roof. If you previously fed the bees, remove the feeder carefully to avoid spilling any remaining syrup and remove the crown board. Give one more puff of smoke and take out the outside frame, using the hive tool to 'crack' the propolised areas where the frame sticks to its neighbour. Hold the frame up level, hanging vertically, and note the contents. In all inspections you are looking for:

a Presence of queen: you might see her if she is marked, but at this time the presence of eggs is good enough. If you have eggs you can be sure a queen was there and laying within the last three days.

b Presence of food: there should be some cells filled with nectar or honey for the bees' immediate use.

c Presence of pollen: without pollen the bees cannot rear brood.

d Presence of brood of all ages: in the case of the nuc', there should be brood of all ages somewhere on the frames, probably in the centre ones. This means eggs, larvae, and sealed cells with possibly some young bees chewing their way out. In the case of a swarm, there should be eggs and larvae present but the larvae may not be old enough to be sealed over.

All this presupposes that you have better than average eyesight, can spot an egg and can readily tell the difference between sealed brood and sealed honey. To help you spot eggs the best advice is to turn so that the light comes from over your shoulder, and even then it is not easy. Spotting pollen is easier as it is usually coloured differently from the cells holding it. Sealed worker brood has a fabric-like surface as opposed to the pure waxy finish of sealed stores. There is no simple way of learning these

differences. Ask your beekeeping friend to go through the frames with you, but don't keep them out too long in the cold.

Go carefully from one frame to the next looking for these things. In later inspections you will also be looking out for diseases, but at this stage just get to know the look of the frames. Give a gentle puff of smoke each time you remove a frame. Use smooth, slow movements and don't drop, jar or shake the frames. If, in the case of a swarm, you see no eggs, do not panic. Your bees might be queen-less, or your 'swarm' might in fact be a 'cast' with a newly emerged virgin queen who will need time to mate and then come into lay. Just give her another week. (See section on swarming, p. 79.)

Be very careful to replace the frames in exactly the same order as you removed them. Give another puff of smoke and close the hive up, putting everything how it was when you started. Make a note of the date of inspection, and what you found, in a notebook and look again a week later. To encourage expansion keep ahead of the bees with frames of drawn comb or foundation. Always make sure there is at least one frame at the outside without any bees on it. Never be afraid to feed if you think the stores may be low or where all foundation is given. Bees have to consume about five pounds of honey to produce a pound of wax, so they devour a lot of food while drawing out foundation.

Nectar, pollen and further expansion

Further expansion of the nucleus depends to a large extent on the time of the year and current weather. In Britain the bees have to collect in three, or possibly four, months all the nectar they need to survive for a year. There is usually no nectar available in worthwhile quantities until the flowering currant bursts forth, usually towards the end of April.

There is a profusion of flowers in May, a lot less towards the middle of June, and in July, if and when the lime trees secrete, the major nectar gathering time occurs. With the exception of the heathers, there are few flowers which secrete much nectar in August, so to all intents and purposes May, June and July are the nectar gathering months. In June there can be a definite break between the end of the first flow and the start of the major nectar flow. The 'June gap' can cause problems in an expanding colony with insufficient reserves of food, and a small feed is often necessary at this time.

To return to the expanding colony, either swarm or nuc'. It will be seen that expansion depends upon the availability of adequate nectar and, possibly even more important, pollen. Pollen is available in reasonable quantities in most parts of Britain, from the early pollen of the crocus to the late pollen of the ivy. A full colony with lots of brood consumes many pounds of pollen annually. Foraging bees get the pollen grains all over their bodies while sucking out nectar from the plants. The pollen is cleaned from the body, using special comb-like hairs, moistened with nectar to bind it and then stacked in pellets on the bee's hind legs in specially designed arrangements of bristles called 'pollen baskets'. The bee then takes the pollen back to the hive in these baskets.

Inside of hind leg showing pollen basket and pollen comb; bee with pollen load.

The changing brood pattern

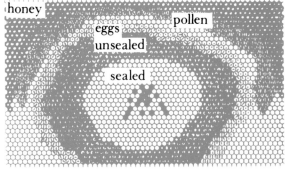

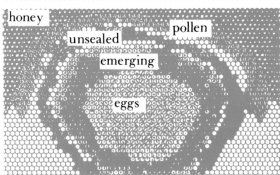

The changing brood pattern (1)

The changing brood pattern (2); note the two queen cells

The changing brood pattern (3)

Inside the hive the bees take the pollen pellets and stack them in cells in the main brood rearing areas so that the pollen is close to where it will be needed. In a healthy frame from an expanding colony the centre areas will be found to contain the brood in rough circles, centred around the cells where the first eggs on that side were laid. A likely pattern will be sealed brood, surrounded in turn by unsealed brood, eggs, and a ring of pollen with an arch of honey reaching to the top bar. As the sealed brood hatches out in the centre, new eggs will be laid and the picture will be partially reversed: two weeks later it will look as first described when these eggs, too, become larvae and then sealed brood.

To return again to your expanding colony. If, on the second weekly inspection, you discover that the queen is laying so well that all but three or four frames have brood on them (even if it is only a few eggs on the outside frames) and there are fairly thick coverings of bees on all but the outside frames, the time has come to add a super. The super should

41

consist, preferably, of drawn combs, although a mixture of drawn comb and foundation will suffice. To 'super up', smoke the front of the hive, remove the roof and crown board, and direct smoke across the tops of the frames to force the bees down into the hive. Then, very carefully, place a queen excluder on top of the brood body, put the super on top of the excluder, replace the crown board, and finally put the roof back. The bees now have a greatly increased area of comb. If there are sufficient bees and sufficient nectar they will store honey in the super, but in any case they will now have sufficient space to avoid overcrowding. Should the expansion continue apace so that next time you look you see bees thick on all but the outside super frames, you may need to add a second super.

For normal routine colony inspections many beekeepers use a 'manipulation cloth'. Although not strictly necessary, I find it very useful for a variety of reasons. A manipulation cloth is merely a length of tough cloth, roughly the same width as the hive and with metal rods or wooden bars to act as weights on

Examining a hive:
(1) Taking off the supers

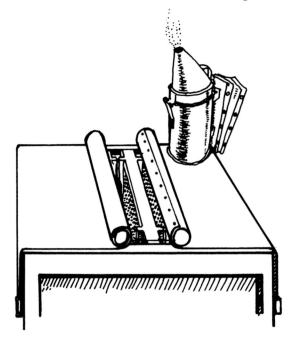

Manipulation cloths

(3) Taking out the end brood frame

42

(2) Gently lifting the queen excluder

(4) A nice brood frame; worker brood in the centre, drone brood at the side

Queen clip and queen

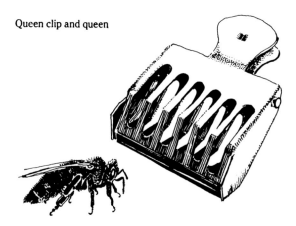

each end. The pukka purchased models have a rectangular gap in between two shorter lengths of cloth. The cloths are used to cover up the top of an open hive, leaving only one or two frames visible. If frames not being inspected are covered up, fewer bees take to the air to bother the beekeeper and no bees from other hives can sneak in to steal honey while the hive is open. My cloths are cheap home-made jobs, old curtain material with thin wooden rods nailed to each end, but I would not like to be without them. I find them useful if I have to break off halfway through an inspection; I merely cover the open hive with the cloth and the bees are safe and contented.

Another piece of equipment which is often very useful is a 'queen clip'. It looks like a device to squash lemons and you might expect it to squash queens, or at least cut their legs off. In fact, it is designed to catch and trap the queen without hurting her in any way. The queen, as we have seen, is vital for the survival of the colony. She can easily be crushed in rough manipulation, leading to a queen-less colony and problems. If the queen is seen and caught in a queen clip early on in a manipulation, she can be placed in a safe position outside the hive and then returned when any rough operations have been completed. The beekeeper never actually touches the queen, so she does not smell strange and the bees accept her back again without fuss. If you purchase a queen clip do find a tin to keep it in as they are fragile and do not work too well after they have been trodden or sat upon.

The summer season is all too short. Unless you find yourself in an area where heather abounds, the honey gathering season finishes, to all intents and purposes, as July turns into August. The expansion which has been taking place in your colony will finish with the nectar. August is a quiet time for the bees. In many cases there are a lot of workers in the colony and little forage outside for them to collect. They conserve stores by staying at home in the hive, which is a good place for them to be since they can be a nuisance at times.

Bee fanning at the hive entrance

One way in which they can cause annoyance or anxiety is in their quest for water. On a very hot day bees cool the hive by fanning. This involves a line of bees standing in the entrance, facing into the hive and furiously fanning their wings. Another row of bees fan just as strongly from inside so that cool air is drawn in and warm air is expelled. If at some point the hive temperature gets too hot for comfort in spite of the fanners, foragers are sent out to collect water. Upon their return the water carriers are met by younger hive bees and the water is taken from them and spread thinly over the inside surfaces of empty cells on the comb. The water evaporates and in evaporating cools the surrounding area. Large quantities of water are needed for this purpose, in addition to water needed for diluting honey which is to be fed to older brood.

Bees in search of water can be less than fussy in their choice of water source. Wet drain covers are a favourite drinking place — not a pleasant thought to a honey consumer, and probably a nuisance to the householder. Once bees become accustomed to visiting a place for water it is difficult to stop them, so a solution is to get in first by contructing some sort of drinking place in a spot convenient to you and the bees. A container with a tap set to drip steadily onto a large stone is one solution. Another idea is a water-filled washing-up bowl with a cloth in the water which hangs over the edges of the bowl. Tubs or buckets of water with floating wooden rafts also work well. Open pools attract bees who promptly drown in them. Siting a drinking place close to the hive seems sensible from the work study point of view, but the bees, usually masters of efficiency, seem to ignore it and go further afield.

If there is little outside activity and you have no good reason for opening up the hive in August you are advised not to do so. With no work to do there will be many bees around just waiting for something to happen, all a bit frustrated at having no nectar to gather. Should you come along and try to tear their home apart they will probably take up the cudgels and have a go at you, and if you do inspect the brood nest you might only worry yourself. The workers, quick to sense the end of the nectar flow, cut down the food to the queen and she cuts down her rate of lay. If you look inside and see no eggs at all you will probably assume your colony is queen-less. This condition, where the queen stops laying completely in the autumn, is quite normal. If the queen was there, or you found eggs, last time you looked just hope for the best and leave well alone.

If you do open up the hive and find yourself surrounded by a host of bees it is entirely possible that they do not all belong to you. In opening up you expose frames of stores to all comers, and bees from another hive can be there in minutes to steal the honey. Some colonies of bees are more aggressive robbers than others, but most will have a go if the opportunity arises. Autumn inspections are definitely occasions when the use of the manipulation cloth pays off. If bees from another hive do manage to steal some honey they will come back for more. Robbing is easy to start and difficult to stop. When you have closed up the hive the robbers will try to dodge the guards at the entrance

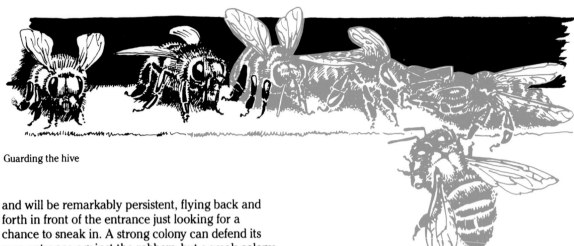

Guarding the hive

and will be remarkably persistent, flying back and forth in front of the entrance just looking for a chance to sneak in. A strong colony can defend its own entrance against the robbers, but a weak colony can be literally overrun by the aggressors. Little groups of bees become engaged in mortal combat all over the entrance area, dead bees litter the front, and the beekeeper sees his colony decimated before his very eyes. Should this happen the first thing to do is to reduce dramatically the size of the entrance. This gives the guard bees much less of an entrance to defend and they can pick off the attackers one by one. A piece of foam rubber or a reduced entrance block both work well, but in some cases it is necessary to restrict the entrance so that only one bee at a time can enter.

Robbing can happen at any time. Smaller colonies, especially nuclei, are much more vulnerable. A colony can literally be left with no food at all, and will starve in a day or two. Be especially aware of this possibility when feeding bees at any time. A small quantity of spilt sugar syrup is sufficient to

unlease the robbing instinct. Robbing is also a way in which bee diseases are spread. Should a colony die out as a result of one of the more serious bee diseases, its stores are then left undefended and up for grabs. A nearby colony will commandeer the stores and take them back to its hive, taking the disease as well.

Wasps can also be a nuisance at this time of year. They change their diet in mid-summer and are then always on the look-out for easy pickings in the honey line. A reduced entrance is the solution to this problem as well, but the pests can also be caught in wasp traps placed near the hive entrance. A jam jar, or a bottle with a jam and water mix, will attract and drown lots of wasps but only the odd bee.

If you started your beekeeping with a swarm or a nucleus you will probably have to defer your first honey taking until next year. The bees will probably not have built up enough to store honey in the supers and it is not done to take any they have stored in the brood body. They need all of that, and more, to survive the winter.

If, however, you caught your swarm early on in the year, or started with a larger than normal nuc', or if you purchased a full colony, it is quite possible that by September you can extract some honey in your first year. For most of us the honey crop is what beekeeping is all about, and the fun and satisfaction of extracting your own honey for the first time makes it all seem worthwhile.

Before starting to extract, however, you need to make plans. Firstly, you have to remove the honey supers from the hive. The bees have laboured for many hours and flown many miles to bring the nectar to the hive and have lovingly guarded it while they turned it into honey. It is hardly surprising that they are not over-pleased when you come along and try to take it from them. A certain cunning is therefore called for to rob the bees of that which they hold most dear without giving them the chance to put up a fight for it.

The nectar gathered by the bees in the first instance is a fairly weak sugary solution, bearing the flavour of the plant which secreted it but containing only twenty to forty per cent of sugar. The foraging bee sucks up the nectar into its honey stomach and in doing so adds an enzyme to it. The enzyme helps in the gradual change from the weak sugar solution to honey, which contains eighty per cent sugar. At the hive, the enzyme-enriched nectar is passed by the forager to a younger house bee who in turn adds more enzyme as she stores the nectar in an open cell. The natural heat of the colony causes water to evaporate from the nectar solution until it reaches over eighty per cent sugar. At this point it can safely be stored by the bees for the winter and each cell is sealed over with wax. This prevents the honey absorbing water and returning to a form in which it can ferment. From the free-flowing liquid form the nectar becomes steadily more viscous until, when sealed, it has the viscosity of treacle. Sealed in the cell the honey will start to crystallise. The source of the nectar determines if this crystallisation, often called granulation, will be a fast or a slow process. With honey from brassicas (mustard or rape for example) it can happen in the comb within a couple of weeks of liquid honey being stored. With most other honeys, lime honey for example, crystallisation is much slower, taking months not weeks.

Once we extract the honey from the comb, however, the crystallisation process speeds up, and any reduction in temperature accelerates the process. For this reason the extraction should be done as soon as possible after taking the honey supers from the bees. If done promptly, the honey is still liquid from the warmth of the hive. It spins out easily and runs through the filters. If the honey stands some time before extraction, and gets cold into the bargain, it is slower to spin out, takes an age to run through a filter and if a few crystals are present these clog the filter and the flow might stop completely. The simple rule is, therefore, to extract as soon as possible after taking honey from the hive.

To check to see if you have any honey to remove takes only a few minutes. Don your protective clothing, light the smoker, smoke the bees, take a short pause, then remove the roof and crown board. Gently smoke the top bars of the super and remove a few frames. If they are all feather-light, with only the odd cell containing unsealed honey, forget it. If, on the other hand, you have a few frames heavy with sealed honey you might decide it is worthwhile taking some, even if only a few pounds. Replace the frames, close up the hive and set about borrowing an extractor.

The extractor is usually the most expensive single piece of equipment needed by the beekeeper. It is also an excellent investment as second-hand extractors many years old change hands regularly at several times their price when new. Unless you have a windfall to spend you are not advised to purchase an extractor straight away. Wait until your second year, ask around and you might hear of a cheap second-hand one. Meanwhile if you need one in a hurry try to borrow one. Many beekeepers' associations have extractors which can be hired or borrowed by members, or your beekeeping friend may lend you one.

The first week in September is the traditional time

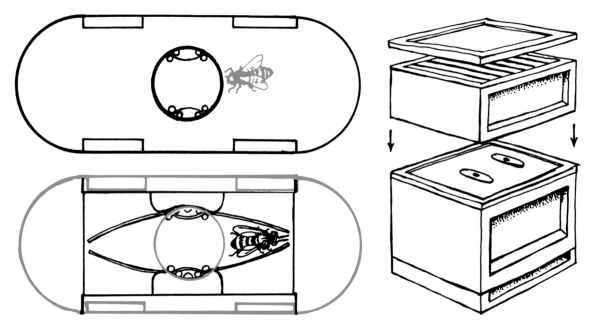

The Porter bee escape and its use on the hive

to extract honey. By that time the honey will be capped over and there should be no danger of extracting any freshly collected nectar in the combs. Any such nectar added to the honey could cause fermentation in the jars. Fermentation causes bubbles of froth which build up pressure inside the jars and force their way out, making a nasty sticky mess on the outside of the jar. The honey smells bad and tastes very peculiar.

Unfortunately, the first week in September will probably be the time everyone else wants to use the extractor, so get your booking in early. Having made arrangements to borrow it on a certain day, you will need to ensure you have the supers available for extraction by inserting a 'clearer board' into the hive twenty-four hours beforehand. I would recommend the use of a clearer board at all times. In most cases a clearer board is the normal double-slotted crown board with a Porter bee escape in each slot. These devices, of which the older models are made in tinplate and the modern ones in plastic, are a sort of bee valve. When placed in the top of the board between the supers and brood body they allow bees to squeeze through a double spring device on their way down to the brood body, but prevent the bees from returning to the super. The clearer board only takes a few minutes to slip into position and about twenty-four hours later you should be able to just lift off the supers and find them clear of all bees. That's the theory. In practice some bees always seem to

want to stick with the stores through thick and thin and make life difficult for you when extracting.

A word of caution here on the use of the clearer board. If it does its work as planned, in a few hours your super should be relatively free of bees, and you will have a large quantity of unprotected honey in the middle of a lot of hungry bees. If there are any bee-sized gaps anywhere above the clearer board the bees will go robbing with a vengeance and all your honey will disappear in no time. This is true of all types of hive, but is especially relevant to the older WBC hives, many of which appear to have gaps at every corner.

If after twenty-four hours the supers are still full of bees, you should leave things as they are for another day since some bees take their time to clear. If they still have not cleared, you should check the Porter bee escapes which might be broken and letting bees up from the brood body, or might be jammed with dead bees so that none can get down at all.

If after twenty-four hours you find only a few bees present, my advice is to take the supers a distance from the hive, then take the frames out one by one, brush the bees off and make them fly back to their hive. For this purpose one can buy a bee brush, but long grass does just as well.

The other way to get rid of a few bees is to take the supers inside with a cover over them, remove the cover and run a vacuum cleaner (cylinder type) over the top bars. It seems a bit unfair to condemn your

48

labourers to a quick and dusty death, but a few bees can be remarkably off-putting for the next part — the extraction.

Honey from comb to jar

The first thing to remember about extracting is that you are into the business of producing food which people are going to eat. This is an obvious statement, yet one which I think needs to be made. Today we have come to expect that our foods are produced in conditions as clean and hygienic as possible, and customers will expect your honey to be of the same high standard. The word 'customer' might seem out of place at this time, since your main aim is probably to produce some honey for your own use, or to give to friends, and you might have no intention at all of selling any honey at this time. There is, however, a strong and steady demand for home-produced honey and the chances are that someone will ask to buy some of yours before too long. If your honey is as carefully prepared as anyone else's, you need have no qualms about selling a few jars.

The extraction of the honey crop is definitely an occasion when help from the family should be encouraged. Several different processes are involved in the transfer of honey from the frame to the jar, and with the help of the family (or friends) it is possible to set up a miniature production line and get the job over more quickly and more efficiently than you could working solo.

Firstly decide where you will do the extracting. The room should be bee-proof, in case bees are still flying outside and are attracted inside by the smell of honey. The room should also be clean, dust free and warm. Ideally, you should be able to work in comfort in your shirt sleeves. Kitchens and garages are often used, but both have disadvantages: honey on the floor seems to make sticky footsteps for days in the kitchen, and honey extracted in the garage can absorb garage-type smells to its detriment.

Whichever room you choose you should cover the floor with newspapers first and then move in the supers containing the honey. For the whole operation you will need:

1 the honey supers
2 the extractor

3 an uncapping knife
4 a bowl to hold the cappings
5 a board upon which to rest the frames during uncapping
6 a coarse-meshed sieve or strainer to sift out wax particles from the extracted honey
7 a finer meshed strainer (nylon mesh or muslin) to take out smaller wax particles (this is not strictly necessary if honey is for your own consumption)
8 a settling tank or ripener, or large container of some sort
9 some jars

A flow-chart showing the processes involved in extraction might look like this:

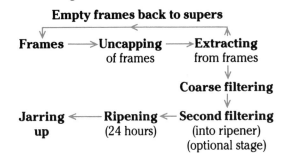

Uncapping

Honey is hygroscopic, that is, it absorbs water. When it does, natural yeasts cause it to ferment. To prevent this, the bees seal over the cells with a capping of wax. It is not possible to extract honey while the cappings are on the comb, so the first operation is to remove them. Special uncapping knives (and forks) can be purchased, but for the small-scale honey producer a bread knife works reasonably well.

The frames are taken one by one and lodged on a board over the top of a bowl or bucket into which the cappings will drop. The knife is placed on the flat face of the comb and with a sawing motion the blade is moved from one side to the other, cutting off the cappings to the cells one or two millimetres below the surface. If the comb is leant over slightly, the cappings fall straight in the bowl, otherwise they have to be scraped in. A messy job this, so you need a cloth, water and a towel for when the telephone rings! Both sides of the frame need uncapping, and the first side uncapped will begin dripping honey as

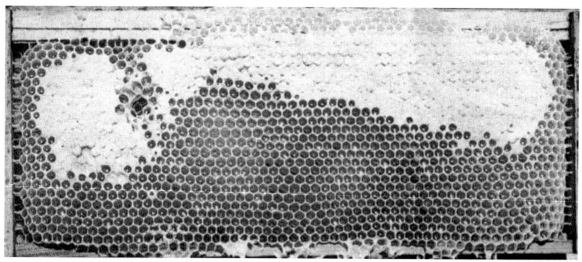

A shallow frame of honey, partly capped

soon as you turn it over to start on the second.
Transfer the uncapped frames straight into the
extractor by the shortest possible, paper-lined route.

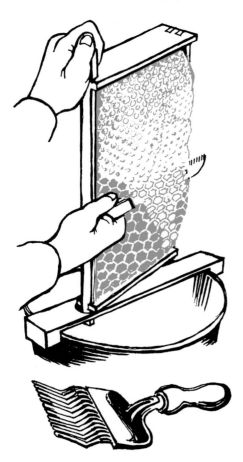

Above: uncapping with a serrated knife; **below:** an uncapping
fork

Extracting

Extractors are basically dustbin-like containers in
which are fixed racks to take a number of frames.
Some extractors only take two frames, larger ones
take two dozen or more. Some have electric motors
to power them, some have a handle and you to
power them. Some are made of plastic, some
stainless steel, some non-rusting alloy and some of
tinned steel which rusts wherever the tin is knocked
off. In all of them there is an arrangement whereby
the racks and the frames inside are rotated at speed.
The honey is thrown out by centrifugal force, hits
the walls of the container, runs down and collects at
the bottom. A tap is usually fixed somewhere near
the base so that the honey can be run off from time
to time.

There are two main types of extractors, radial and
tangential. In a radial extractor the frames are
loaded in such a way that the line of the comb forms
a radius to the container, i.e. the comb starts at the
centre and stretches to the circumference of the
extractor. In the tangential models the frames are
arranged so that the comb is at right angles to the
radius, in this case stretching from one point on the
circumference to another.

Most larger models work on the radial principle,

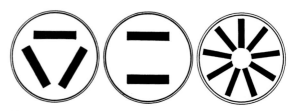

Extractors
Left to right: 3 frame tangential, 2 frame tangential, 9 frame
radial

50

A radial and a tangential extractor

of supers a radial extractor has many advantages. The frames are loaded, spun until the honey flies out, and then spun progressively faster until no more honey is seen to splatter on the inside walls of the extractor.

In all cases you are advised to load the frames in such a way that they are balanced in weight inside the extractor. This is necessary as some frames are light and almost empty, while others are brimming with honey and heavy as lead. An out of balance situation causes the whole extractor to vibrate, and at speed it starts to move round the room in an alarming way.

There is usually a honey collecting space beneath the rotating frames and a tap to draw the honey off. Keep an eye on the honey level and run some off before the top surface of the honey is stirred by the top bars of the frames. This is one point where you definitely need help, even if you are working alone, since the tap is close to the ground and must be raised sufficiently high to get both coarse filter and receptacle under it. To attempt this by oneself is to court disaster and risk rupture.

Coarse straining

Through the run off tap will come honey, some pollen, a mass of bits of wax from the remains of cappings and, on occasions, parts of dismembered bees. All these foreign bodies must be filtered out and the best plan is to take out the largest pieces straight from the tap. For this purpose a fairly coarse filter of some sort is required. Special purpose wire mesh filters (tap strainers), which hang from the tap, can be purchased, but a plastic or wire mesh kitchen sieve supported in some way over the top of a plastic bucket will suffice. A problem at this stage can be a honey flow which is reduced rapidly to a trickle as big pieces of wax become clogged in the holes in the mesh. An occasional stir with a wooden spoon will help by keeping the wax in suspension.

At this point you have to decide how clean you want your jars of honey to be. After the coarse filtering there will still be small pieces of wax and other unidentified solid specks in the honey. If you intend to sell any honey these must be removed; if you intend to consume the honey yourself you will probably be happy to eat it straight from the coarse

while most small, hand powered ones are the tangential type. The chances are that if you borrow an extractor it will be a tangential model. Tangential extractors are more difficult to operate than radial ones; the uncapped open cells on the outside of the comb point straight at the walls of the extractor and the honey flies out at a comparatively slow rate of rotation. Meanwhile, the honey in the cells on the inside is pushed by centrifugal force into the bottom of the cells and when the outside honey disappears the comb tends to break up with the pressure from the inside cells. To avoid this, the frames have to be turned twice during each extraction; a sticky and time-consuming job. The full sequence for *tangential extractors* is as follows:

1 load frames
2 rotate gently
3 reverse frames
4 rotate gently
5 rotate fast
6 reverse frames
7 rotate fast

With one or two supers of honey this double rotation is no problem, but for extracting a number

Two coarse strainers, strainer cloth and honey ripener

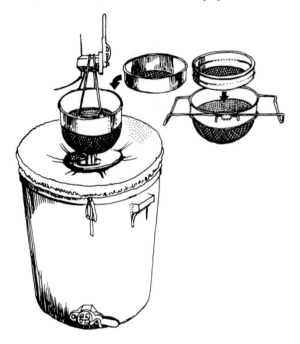

filtering stage, but specks of dirt tend to sink to the bottom of the jar where they are easily visible and pieces of wax can float to the surface and form an unsightly scum.

Optional second filtering

If a second filtering is to be carried out it should be through a nylon mesh, mutton cloth, or similar material. After this second straining the honey has to be left to stand for up to twenty-four hours. Air bubbles are beaten into the honey during the extraction and these rise to the surface slowly and form a froth as long as the honey is kept from becoming cold. With the extractor you might be able to buy or borrow a honey settling tank, sometimes called a honey ripener. These vessels are tall compared to their girth and have a tap near the base. If you can borrow one of these, drape the straining cloth over the open top and tie a length of string tightly round just below the rim of the tank to keep the cloth stretched taut across the top. Make a slight depression in the tightly stretched straining

cloth and pour in the honey. If everything went well and the honey is still warm from the bees this straining should be very quick. Any straining cloth gets clogged with the strained out solid particles after a while, but one cloth should do for at least one or two supers of honey.

If the stuff just will not go through the cloth, I suggest you skip this second stage of filtering for your first venture. The chances are that the honey has become cold and more viscous; honey will not run through a filter in that condition. Gentle heat is the answer, but the heating of honey is a subject on its own. I suggest you settle for less clear honey for your first extraction and try at a later stage to obtain a perfectly clear sample.

Ripening

In the past honey was left for some days in the ripening tank in a warm room. The air bubbles rose to the surface in a day or so, the denser honey sank to the bottom and the less dense honey, that containing most water, rose to the top where it ripened by the evaporation of some of its water. This was a slow process and these days honey tanks, or settling tanks, have replaced ripeners, although there are still a lot around.

As previously stated, any air bubbles in the honey should rise to the surface within twenty-four hours so that when you bottle the honey it will be free of air bubbles.

Jarring

The last and most rewarding stage — open the tap of the honey tank and out pours the liquid gold. If you intend to sell some of your honey you are advised to use the standard types of honey jars which hold one pound, or half a pound, and can be purchased from the equipment suppliers. For your own use you can use any containers available, but remember that other people will judge your honey largely by its looks, so your aim should be clean and sparkling honey in containers to match.

Cleaning up

Before you have the joy of pulling your first pound of honey you will have the less enjoyable task of cleaning up after the extraction.

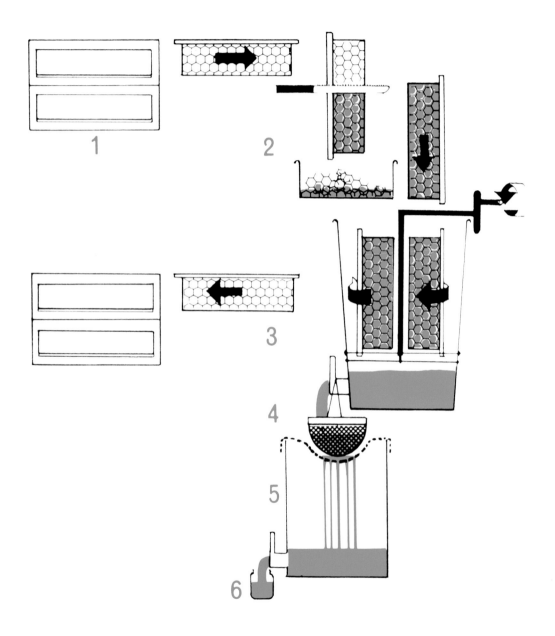

Extracting your honey 1. Full supers 2. Uncapping 3. Extracting
4. First straining 5. Second straining 6. Jarring after 24 hours

Cappings straining into a bowl

and leave the supers another couple of days.

The cappings
You now have a container with honey and cappings, probably standing in the middle of a sea of honey drips. In some way or another you need to separate the honey from the beeswax. One possibility is to filter the honey through the same coarse filter you have already used. Another suggestion is to tie the cappings and the wax from the first filter into a cloth, or the second filter, knot the top and hang up over a bowl to drip. This is easy but takes up to a day to accomplish and the windows must be kept closed for that time or the bees will be in to investigate. Another solution is to place the cappings in an enamel or stainless steel saucepan and to put this in an oven in which a meal has just been cooked. The wax will melt and float to the top, and settle as a cake when cold. The honey beneath will have lost something in the heating but will be perfectly satisfactory for cooking. Do not, however, heat the oven to help this process as beeswax can vaporise and the vapour is highly inflammable.

Another way of dealing with cappings is to wash them in cold water, strain out the wax, dry them on newspapers and use the sugary water either to make mead or to use in making up winter feed for the bees.

The wax, either in a cake from the oven or in flakes from the washing, should be dried and stored in an airtight container or plastic bag. Beeswax is a valuable by-product of beekeeping and is always a marketable commodity.

The supers of wet frames
These will not be wanted until next spring. The frames can be stored wet or dry (see storage of supers, p. 60). If you decide to store them wet for the winter, simply move them temporarily to a convenient place and leave with bee-proof covers of wood or plastic on the top and bottom of the supers. If you decide to store the frames dry they must be cleaned up. The bees do this quickly and efficiently for you. In the late evening when bees have finished flying, put the supers back on the hive. Take off the roof, remove the Porter bee escapes from the clearer board, put the supers of wet frames on top of the clearer board and replace the roof. The bees will come up through the open holes, clean up every vestige of honey and take it down into the brood body. After a few days the frames will have been thoroughly cleaned and can be removed. If you go late at night you can often find them clear of bees and simply take them straight off; otherwise, put on gloves and veil and replace the Porter bee escapes

Ripeners, tanks, extractors and sieves
These should be taken outside and washed down with copious quantities of water. Use a hose pipe if you have one. If you use hot water the pieces of beeswax melt and stick to the sides of the container, so use cold water first, then rinse out with hot water and turn upside down to drain. Wash the knife, sieves, buckets and boards in the same way, throw out the newspapers and breathe a sigh of relief, then go back and wipe the honey off the floor, door, windows and cat. Never mind — it is bound to go better next year, and just look at those jars of your own glorious home produced honey.

Honey in the comb

One of the nicest ways to eat honey, and one enjoyed by many enthusiasts, is in the comb. The connoisseur's way is from a honey section, a special thin wooden frame which the bees are encouraged to fill with honey and cap over. The sections are usually placed on the hive in special section racks, but extra-wide frames holding three or four such sections can be hung in ordinary supers. To produce good sections the beekeeper needs a strong colony, a good flow of honey and a lot of skill. Some bees appear to draw out and fill sections and some do not, and those that decline to oblige might just swarm instead!

Beginners wishing to obtain cut-comb honey can do so with ease. Normal super frames are fitted with unwired thin foundation, and when drawn out and capped, pieces of comb are simply cut out for eating, either with a sharp pointed knife or with a Price comb cutter. The latter device cuts the comb out to the exact size to fit into cut-comb containers, which can be purchased from most equipment manufacturers. These special plastic tubs have a thin transparent lid to show off the surface of the cut comb to good advantage. Full sheets of thin foundation are not essential for cut-comb honey. Thin strips of foundation only one inch (25mm) deep can be put in the frame to start the bees producing comb in the right place and they will, with luck, carry on and fill the whole frame.

Beeswax

Beeswax is a valuable by-product of a beekeeping venture. In fact, it usually fetches more per pound than honey. For generations beeswax candles were used by the rich to give a clear smokeless light and were customarily used in churches where protection of the surface of pictures and painted ceilings was important. Tailors wax their thread with beeswax to prevent it knotting, and beeswax furniture polish when applied over a period of time produces a sheen which cannot be bettered. Beeswax also finds its way into other commercial products, from ointments to bubble gum, so there is always a ready demand for any surplus you may produce. Some beekeeping appliance manufacturers will purchase beeswax to offset the value against future purchases. Most beeswax sold back in this way ends up as foundation and the whole process is repeated.

Many beekeepers develop an interest in wax and make their own candles, either by the traditional dipping method or with moulds. Other beekeepers turn their wax into foundation themselves with the use of one of the various small-scale foundation presses. There are classes at most shows for perfectly shaped and cast cakes of beeswax, and some enthusiasts fashion wax into the shape of fruit, flowers, birds and animals in a truly amazing way.

Wax extraction

A brood frame newly drawn out from foundation is usually light in colour and light in weight. After some years the frames tend to become very heavy with a build up of cocoon skins and propolis, and become unsuitable for further use. When they are replaced the old comb can be thrown out or burnt, or an attempt can be made to recover some of the wax still in it. At about 145°F (62°C) beeswax melts, becomes liquid and runs. One can heat the old comb to this temperature on a warm sunny day by making an insulated box, covering it with two layers of glass, and setting it at an angle to catch the sun's rays. If the comb is placed on a sloping metal tray with a perforated metal grid to keep the old comb in place, the wax will (when the sun shines) become liquid and run out into a bowl of some sort, which can be positioned below the tray. The wax sets and at night the solidified cake of wax can usually be knocked out by upturning the bowl and giving the bottom a sharp bang.

Solar-powered wax extractors such as this can be purchased, but the enthusiastic do-it-yourselfer here again comes into his own. I made one entirely from second-hand materials — a pair of old windows screwed together provided a cheap double glazed lid. I cut a packing case to fit, lined it with thin plywood, hardboard and insulation board, and used a large baking tray for holding the comb and a loaf tin to catch wax. Such re-cycling should appeal to conservationists, as should the use of the sun as a source of power. The amount of wax recovered is usually slightly disappointing. One needs a lot of old brood comb to produce a pound of wax. Cappings can also be rendered down in the solar wax extractor to give a more worthwhile quantity of wax. After draining off the honey, the cappings should be washed (preferably in rain water), dried on a newspaper and then put into the extractor. The washing is necessary because when heated in a solar wax extractor any honey in the cappings runs out, too, sinks beneath the wax and turns into a sticky burnt substance like black treacle.

Heather honey

In most parts of Britain the flow of nectar falls off dramatically during August, yet for most of the month many high moorland areas are covered with the purple flowers of the ling heather (calluna vulgaris) which is a most prolific source of nectar. Many beekeepers find it worthwhile to prepare colonies and send them off on their holidays to the moors. The bees should be in position on or about 'the glorious twelfth' when grouse shooting starts, and are usually removed about a month later. Colonies with lots of bees headed by a queen newly reared in May or June appear to do best. Heather honey is thixotropic, which means it is stiff until stirred when it will run freely for a while. It cannot easily be extracted by normal means and the most usual way of coping with the honey is to scrape the frames down to their midribs and press the resulting mushy mess of honey and wax through cloth in a heather-honey press.

Good weather on high moorland areas is by no means guaranteed and most colonies are sent up to the moors with at least one good super frame of stores to ensure the bees do not starve during a sudden spell of rain or high winds, or both.

August is often a hot month and to prevent the bees overheating on the journey the hives are usually closed up in the evening, a travelling screen is added, and the move made overnight so that the hives can be opened up on the moors at first light — a joyous sight for bees and beekeepers alike.

The trek to the moors and the special equipment required make it unlikely that the beginner will consider trying for heather honey in the first year or two, but local beekeepers' groups often combine to share transport and equipment and the run to the moors is very definitely a beekeeping experience to be sampled and appreciated as soon as possible. The whole exercise is fraught with opportunities for mishap and near disaster and stories of spectacular and humorous accidents do the rounds for years afterwards.

Honey for Sale!

Beekeepers supplying honey for sale to the public must obviously ensure that their product is as clean and well presented as possible. HM Government lay down stringent regulations about the conditions for preparation and sale of food products. Beekeepers intending to sell honey are advised to check the latest requirements for producing, labelling, and lot numbering. These regulations change from time to time and details of current requirements can be obtained from your local Environmantal Health Officer.

An easier and more user friendly way is to request a copy of BBKA Advisory Leaflet 103 'So you wish to sell honey?' An additional tip is to purchase any honey labels required from a supplier of beekeeping equipment who will ensure that the labels comply with current legislation.

If, in your first year, your bees brought in enough honey for you to take some for yourself, you did very well. In any case you must now make sure that they have enough stores to live on during the winter or they will not be alive to collect any the following summer.

In the autumn the activities inside the hive are very different from those of a few months earlier: the number of workers will be much reduced and will diminish daily, there will be no drones present, and the queen will probably be laying very few eggs or have gone out of lay altogether. There is, therefore, only a small amount of brood present and the workers now concentrate on filling all the empty cells in the brood combs with stores to see them through the winter.

The bees' aim is to go into the winter with every bit of available comb filled with stores and neatly capped over to prevent it from absorbing moisture. There is little forage available in the autumn, so you will have to supplement what is available by feeding the bees with sugar syrup. Use ordinary white granulated sugar, not brown or demerara which cause digestive problems to the bees. You should buy about eleven one-kilo bags per colony in preparation. You might not need them all, and it is unlikely you will need more unless you have very prolific bees in large hives.

The other piece of equipment you need is a feeder of some sort. These come in all shapes and sizes. In most cases the sugar is dissolved in water and the resulting syrup is placed in a feeder on the hive where the bees can collect it and take it back to store in the brood body. Basically there are two sorts: contact feeders and trough feeders. Contact feeders are containers with tightly fitting lids in which are set pieces of wire gauze, or in which some two dozen small holes have been made. The feeder is filled with syrup and the container is inverted and placed over the feed hole on the crown board. The syrupy liquid cannot rush out as air cannot enter to replace it, and droplets ooze down through the small holes or gauze for the bees to suck up and make off with. Contact feeders can be made at home from hard plastic ice-cream containers, lunch boxes or any tins with push-on lids.

In trough feeders of the Ashforth, Miller or 'Rapid'

feeder type, the syrup is put in an open trough and the bees can enter, after overcoming a simple obstacle course, suck up the syrup and return with it to the brood body. In the case of Ashforth and Miller feeders, the correct size feeder for the type of hive is needed and it is placed on the hive beneath the roof. In most other feeders some extra height must be provided between the crown board and roof to provide clearance for the feeder. One way of doing this is to use an 'eke'. An eke is merely a topless and bottomless box whose sides are the same length and width as the hive and of sufficient depth to cover the feeder when the roof of the hive is replaced. To make an eke is an ideal DIY job, since at its simplest it is only four pieces of wood butt-jointed and nailed at the corners. An empty super works just as well but then you have to find somewhere to put the frames normally contained in it. Frames are fragile things, very easily become damaged, and are definitely better stored in the supers than stored anywhere else.

For autumn feeding a mixture of one pint (½l) of water to two pounds (1kg) of sugar is used. You can measure it out, but the following ploy is easier and will give the correct mix. Fill the container with sugar, shake the sugar level, note or mark where the top of the sugar comes to, and fill up the container with hot water until this level is reached, continuously stirring first to clear air bubbles and then to help the sugar dissolve. After quite a bit of stirring, when no more grains of sugar are present, allow the syrup to cool down before giving it to the bees.

Feeding should start any time after the honey supers are removed. In Britain the suggested time is the last half of September, with the proviso that feeding should be finished by the first week in October. There is a good reason for this: if bees are fed sugar syrup later in the year they will have insufficient time to evaporate the excess water, and the syrup will be stored uncapped and could ferment, causing digestive troubles to the bees in the depth of the winter.

The syrup should be fed to the bees during the late evening. The sudden availability of syrup causes great excitement in the hive and this excitement can spread to other hives and robbing can break out.

Feeding the syrup after flying has ceased for the day causes the minimum of disturbance. (It is a good plan to put in a reduced entrance block whenever you feed to guard against the chance of robbing.) If you are fortunate enough to have a Miller or Ashforth feeder first remove the crown board, put the feeder into position, pour in the syrup, put on the crown board and replace the roof. If you have a contact feeder, remove the roof, put the eke or empty super on top of the crown board, quickly invert the feeder and place over the feed hole before replacing the roof. For any of these operations, veil and gloves are recommended but a smoker is not, although a torch sometimes is!

My suggestion is that you feed each colony as much syrup as it will take. The quantity taken depends, of course, on the amount of stores present at the start of feeding, but I have found that my bees take about two and a half gallons (about eleven litres) of syrup and then just stop taking any more down. Very prolific bees in the largest hives will almost certainly require much more than this and you might then need to buy several more one-kilo bags!

When the bees have taken all they require, remove the feeder, and the eke if used, replace the roof, and feeding is complete. An expensive operation perhaps, but bees will not waste food, and the beekeeper who feeds his bees well for the winter will see them come through in excellent condition and poised to take advantage of the early spring nectar flow.

Feeders

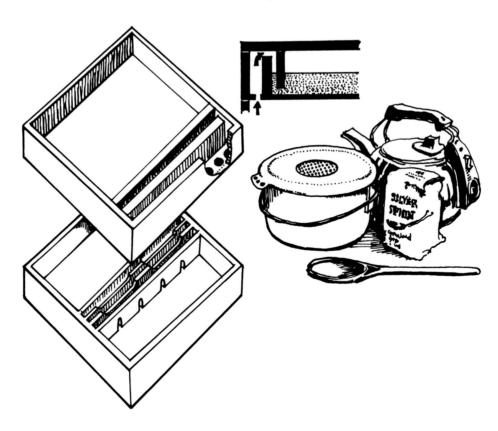

Once the bees have been given the food to sustain them through the winter there is little else you need to do until next year. An hour or two on tidying up and battening down is all that is required.

Mouse guards

During the winter the bees cluster together tightly for warmth, leaving some frames uncovered and unguarded. Mice often steal into the hive during the bees' inactive period, eat the honey and build nests in the comb. The bees only eject these uninvited guests in the spring, by which time the combs are ruined. A mouse guard is the answer. This is a strip of metal with holes of ⅜ inch (9½mm) diameter drilled or punched into it. The guard is fixed with tacks or drawing pins across the whole entrance in October and removed when the spring weather arrives. The holes allow the bees to pass through at will, prevent mice gaining entry, and allow air to circulate freely inside the hive. An entrance block which reduces the entrance to a slot no more than ¼ inch (6.5mm) high also keeps out mice, but unless the central slot is very wide it can restrict the circulation of air to the outside frames and cause these to go mouldy by spring.

Storing supers

Mice are also fond of building their nests in the frames of supers stored for the winter in garden sheds. As supers of drawn combs are valued possessions of any beekeeper, they should be well stored and protected from unwanted visitors.

As previously stated (see p. 54), frames may be stored wet, straight from the extractor, or dry, after the bees have cleaned them up. Either way, I suggest you stack the supers on top of a flat mouse-proof board, drop a few crystals of PDB (Paradichlorobenzene) among the frames, place a newspaper between each super and cover the stack with another flat, mouse-tight board. The PDB crystals, a mothball-like substance, are a guard against damage to the comb by wax moths.

Greater and lesser wax moths

Both of these appear where bees are kept and wax is available. The main difference between the two species is, as the name suggests, in size. With outstretched wings the greater wax moth would effectively cover a 2p coin, while the lesser one would cover a 1p coin. The adult wax moths sneak into the hive, lay their eggs in the honeycomb and the larvae, which hatch out, tunnel along the cappings of sealed brood, eating the wax and leaving a meandering trail of silky strands behind them. The larvae then find odd corners of the hive, tuck themselves away, pupate and emerge as adult moths.

Some bees are better than others in this respect, but most colonies of bees will not tolerate the intrusion of the moths into their brood area and

A mouseguard in position at the entrance of a hive

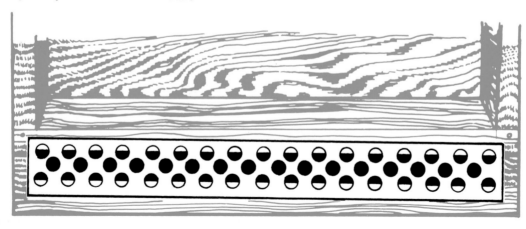

keep them at bay. Stored frames are another matter. Being unprotected by bees, the moths have a field day. Wax moths only have a four-week breeding cycle in summer and can turn good comb into dusty shreds in a short time. PDB destroys the larvae, so all stored frames should be protected by sprinkling with a few crystals. Wet stored frames are less likely to be damaged by wax moths. Stored brood frames are damaged more than super frames, since the moth larvae, in addition to eating wax, also appear to need to eat old cocoon case from cells in which the bees have bred otherwise larval development is impaired.

Hive ventilation
Your hive, now reduced to a single brood body in size, with a mouse guard or reduced entrance block in place, will provide adequate conditions for the bees during winter. However, many beekeepers consider that ventilation in the hive is not perfect with this arrangement. It is said that the air entering the front of the hive flows around the cluster of bees in the winter months and causes draughts on its passage out through the feed hole. A better arrangement is to remove the roof, gently raise the crown board and insert a matchstick under each corner. Completely cover the feed hole or holes with

Matchsticks under the corners of the crown board

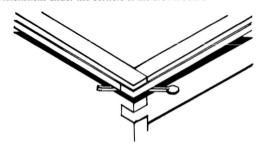

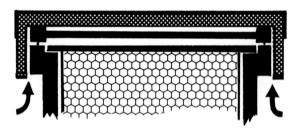

a small piece of board, plastic or glass and replace the roof. Air entering the hive now rises to the top of the brood body, is unable to funnel through the feed holes, spreads out in all directions and escapes through the thin gap, one matchstick wide, between the top of the brood body and the underside of the crown board. In the summer this gap would quickly be filled with propolis, but if the operation is carried out in late October no propolising is likely until next spring.

Weatherproofing hives
At this time of the year it is a good idea to protect the wood of the hives against the coming winter. I paint the whole hive, with the bees inside it, with a foul mixture of equal parts of paraffin, creosote and waste engine oil. This mixture is cheap and possibly harmful to some of the bees, but I have never noticed any sudden losses after its use. Better quality, more expensive products can be obtained, which are stated to be non-toxic to bees. Best quality hives of western red cedar should need no proofing at all, but hives made of ordinary deal need protecting at least once every year if they are to last long. Should big gaps or knot holes open up in the hives they should be plugged before winter. The black bituminous body sealants used on the undersides of cars are suitable for this, as are the tubes of masonry sealant used for caulking joints round doors and windows.

Strong winds can occasionally whip the roof from a hive. The traditional way to prevent this is to find a brick (two bricks in the case of WBC users) and to set the brick (or bricks) on the roof: your bees are now battened down for the winter.

Varroa

The above cosy conclusion to preparations for Winter would have been sufficient until the early 90s when varroa reared its ugly head. No extra actions are necessary in those fortunate areas where varroa has yet to arrive. However for the majority of beekeepers who have colonies in a varroa infected area some sort of detection or control action is probably required after the honey crop has been taken. Varroa is covered on pages 102 and 103, with more details of relevant reading material, and some control advice, on page 109.

The activities of the bees within the hive lessen as the days shorten. On nice days some bees will fly in search of pollen from late flowering plants, such as michaelmas daisies and ivy, but there is little or no nectar to be gathered and the bees, economical as always, conserve their stores by not wasting excess energy in any pointless activity. In most cases, breeding will have ceased completely. No new bees will be emerging to replace those which perish, but because of reduced activity the life span of the workers is dramatically increased. Most of those workers alive in the autumn survive till early spring, so that the hive population will diminish only slowly.

Bees do not hibernate in winter, they merely slow down their activities to a point where they are nearly comatose. In this condition they need very little food to keep them alive and rely to a large extent on body fat put on during the summer, which means that in mild winter conditions food consumption is very low.

With the first frosts of winter, the bees form a cluster. This is not unlike an oval swarm hanging on

The winter cluster

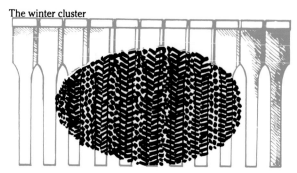

the frames within the hive. The bees pack tightly together into a rugby-football shaped mass. Somewhere in the middle is the queen. Some bees consume food and give off heat in the process to maintain the required temperature in the centre of the cluster. The bees on the outside become colder than those on the inside and over a period of time they rotate positions. The colder the weather, the more food is needed to maintain the temperature. The position of the cluster changes as the bees consume stores in one part of the hive and slowly move on to another.

During these cold spells the bees cannot emerge for defaecation flights, sometimes for weeks at a time; but, they are clean creatures who will not willingly foul the hive, so their bodies have become adapted to storing waste matter until the weather improves and flying is again possible. After such an enforced confinement any sudden improvement in the weather will result in the air round the hive being full of flying bees. A lovely sight and reassuring to the beekeeper, but remember the primary reason for the sudden activity and do not go too close or you might win one or two unsolicited presents.

The bees can only survive in cold weather by clustering together within the hives to conserve heat. Some occasionally venture forth outside during warmer spells of weather to collect water which they then use to dilute the honey. When the ground is covered with snow and the sun is shining the reflected light attracts bees out in large numbers. They are safe while they remain flying, but as soon as they settle on the snow they quickly lose heat to a point at which they are unable to fly again and so

they perish. Large numbers of bees can die in this way. The remedy is for the beekeeper to shield the reflected light entering the hive by resting a large board or sheet of metal on the front of the hive to obscure the entrance. Occasionally the snow falls deeply enough to cover the entrance completely and needs clearing away to allow proper hive ventilation.

Any bees which die in the hive in warmer weather are dumped unceremoniously out of the entrance. In the winter they fall to the hive floor where they remain until a warm flying day when a large number can be thrown out in a few hours. This is perfectly normal; it does not mean your bees are necessarily suffering from any of the diseases which can afflict them.

After several weeks of bad weather it is possible for the entrance to the hive to be almost completely blocked inside by dead bees. A clear space must be left for easy access of both bees and fresh air, so it is sometimes necessary to scrape out dead bees from the floor boards. A long bladed screwdriver is adequate for this operation.

Around the time of Christmas and New Year the queen is given more food and will lay a few eggs. The outside air temperature can be hovering around zero but the bees, by clustering more tightly, will raise the temperature in the area of the eggs to about 93°F (34°C). The eggs will hatch and the larvae will develop into pupae and then adult workers in the normal way. From then on, a few more eggs are laid each day so that hive numbers start to increase steadily from New Year.

Some beekeepers give their bees a Christmas present in the form of a cake of candy. This is white sugar in the form of bakers' fondant. It can be made at home or purchased from equipment suppliers. A cake of candy is placed over the feed hole and if the bees are hungry they will dilute the candy with water and use it for feed. If they are not hungry they may ignore it completely. Some people use candy instead of sugar syrup for autumn feeding. It is not recommended for this purpose, however, since the bees have to fly out in search of lots of water for dilution purposes, but fed in an emergency to a colony close to starvation it is a good bet. If all went well with your autumn feeding and you fed your bees as much sugar as they would take down you

should have no fears for their starvation and no need to worry about topping up the stores with a feed of candy.

As the days lengthen in the New Year some early bee plants come into flower. If the weather is favourable, the bees will be out in force and can be seen almost tearing apart the early crocus flowers in their quest for pollen. Although last year's pollen will still be available in the hive (the bees store it in the autumn, covering it with honey and capping it over in the normal way) it does seem that bees prefer to collect new season's pollen to feed to the larvae. The sight of the bees landing at the entrance laden with pollen is a sight to gladden any beekeeper's heart. It tells the beekeeper that brood is being reared, and he knows that the availability of a good supply of early pollen for brood rearing is the first requirement for obtaining a bumper honey harvest.

Be prepared

The months from October to March are quiet months for the beekeeper. For the most part the bees will look after themselves quite well; in fact, they are best left alone. Even lifting the crown board and taking a quick peep at the top of the brood frames will alter the temperature in the hive and possibly cause the bees to consume more food later to restore the temperature.

During these quiet months there are many things you, as a beekeeper, can do to prepare yourself for the next season's activities.

My first suggestion is that you enquire about evening classes on beekeeping in your area. In many places prominent local beekeepers organise classes for beginners, which can be held anywhere from village halls to technical colleges. By attending a

A beginner, frankly speaking, would require more protection than this when examining bees!

course of lectures you will greatly widen your beekeeping experience, both through the formal tuition and the informal contact with other beekeepers. We all have problems and make mistakes when learning anything new — beekeeping is no exception. You will feel much better about your own problems when you hear the messes other people get themselves into. Public libraries seem to be good places at which to inquire about evening classes. Some counties actually have a

county beekeeping specialist who might be called an adviser, officer, lecturer or instructor. He may be full-time or part-time, but if your county is fortunate enough to have one he will almost certainly know all about local courses, even if he is not taking them himself.

As well as attending formal evening classes, you should also consider joining your local beekeepers' association. If you do, you will discover that there is no such thing as a typical beekeeper. They come in all shapes and sizes, both sexes, all ages and all occupations from stockbroker to schoolboy. Some members will have been keeping bees for years and some for only weeks. All beekeepers seem to like to talk about their bees and you will quickly discover that your limited experience does not put you at a disadvantage. Some beekeepers appear to have kept bees for years with only a hazy idea of what is happening in the bottom half of the hive.

Some local associations function at a purely beekeeping level, while some have an active social whirl from formal dinners to tea on the lawn. During the summer months most associations organise bee handling demonstrations of some sort where competent local experts perform colony manipulations, explaining their actions as they proceed. These are excellent value and provide an opportunity to see how others go about it and ask endless questions about your own bees as well as the demonstration colony.

There are often special insurance schemes open to association members and usually a well established grapevine system from which you can get to hear of second-hand equipment which might be for sale. The organisation of local associations takes time and effort. Do not let on that you can read and write or you will be on some committee in no time at all, and will quickly become deeply involved. I made this mistake myself, but can honestly say that I have had a great deal of enjoyment and satisfaction from this involvement.

One of my earliest beekeeping acquaintances was the late Frank Richardson, a life-time beekeeper and acknowledged master of the craft. For ten years he and I together were concerned with organising five-day residential beekeeping courses at our local college of agriculture. An intensive summer school course such as this seems to me to be the very best way to find out if beekeeping is for you – but then I am biased. We have had beginners who could not tell a bee from a wasp on Monday handling colonies unaided, and pointing out the queen, by the end of the course.

There are still many opportunities for attending either longer residential or one-day courses on both basic beekeeping and more advanced specialist subjects. Details of such courses are to be found in specialist beekeeping magazines.

The most widely read magazine is *Beecraft,* the official journal of the British Beekeepers' Association. *Beecraft* offers a wide range of articles of interest to beekeepers, information on past and forthcoming beekeeping activities, book reviews, items for sale, and a mass of addressees and contacts of all sorts. Some local associations have schemes whereby members obtain *Beecraft* at special rates as part of an annual subscription.

Most county beekeepers' associations are member associations of the British Beekeepers' Association. BBKA publish a quarterly news letter giving the latest association news. This is often available to members through their local association and is well worth reading.

One leading equipment manufacturer, E.H. Thorne Ltd of Wragby, publish their own quarterly magazine, *The Beekeepers' News,* which they will send to any beekeeper for the cost of postage and packing.

Current address for obtaining further information about these magazines are given on page 108.

Spare equipment

As soon as you are dealing with full colonies in the summer months, swarming and swarm control will be subjects of major concern. The chances are that some time from May to July your colony will make

preparations to swarm and you will need equipment to deal with this. The slack season is the best time to prepare. You will need a hive of the same type as the one your bees are occupying, ten or so spare brood frames and foundation for each. By asking around you may hear of a cheap second-hand hive. New frames often come in the flat and need making up — a tedious business and one which is best done at leisure rather than in a rush when they are needed that minute. Foundation can be slotted in the frames as they are assembled, before fixing the second bottom bar, or the frames can be made up first and the foundation slipped in later. Foundation can become brittle, and is easily cracked in cold weather. I find it best to warm it in the airing cupboard for a day before handling it. Gentle heat only, though, or you will have a puddle of melted beeswax. Beeswax smells beautiful to most of us; the bees seem to like it, too, and will more readily work fresh smelling foundation. For this reason foundation is best left in an airtight polythene bag until shortly before it is needed.

Spring

This is the season of hope and sometimes starvation. As the days are steadily lengthening, the birds sing earlier each day, spring flowers appear, buds break out on seemingly lifeless twigs and the bees in their hive are already planning for their sudden spring expansion. In marginal weather conditions there may be little outside activity; the bees inside are still clustered together, yet their numbers are increasing daily.

Then one day in March or April the sun will shine and the bees will be out *en masse* enjoying it. If pollen is available from crocus or willow, they will be taking it home as fast as they can. Inside the hive the cluster will have broken and the bees will be extremely busy cleaning and tidying up every nook and cranny.

This is the moment for the beekeeper to don his protective clothing and make a first quick inspection of the year. The whole thing should take only a matter of minutes.

First spring inspection

Smoke the hive gently, remove the lid and gently prize off the crown board. Take out matchsticks if these were used for ventilation purposes, since they will not be needed now. The feed hole covers will not be needed either, as feed holes can be left open from now on. If you have manipulation cloths this is the time to use them. By covering up all but the frame being inspected you retain more heat, make the bees' task easier and help them conserve stores. Remove the outside frame and quickly inspect frames from one side of the box to the other, keeping no frame out for long and replacing them in the same position they occupied before. You are looking for two important things: food and brood. Without the food there will probably be little or no brood. Towards spring the remaining stores will be consumed at an ever increasing rate as worker bees, by flying more, consume more food and as the quantity of brood being reared increases.

It is very common for colonies which have survived the winter to eat all their stores and to starve in early spring. It is said that more colonies die out in March than in any other month. If, therefore, you find your bees have some sealed stores on their frames but the total area covered by stores is only the size of a saucer you should finish the inspection, close up and carry out an emergency feeding to bump up the stores. There might be very little brood present in these circumstances as the bees will already have sensed their slim prospects of survival.

Emergency feeding

A cake of candy placed over the feed hole is the easiest solution. If you have one, use it. At this time

of year Ashforth or Miller trough feeders are not suitable: when the temperatures are low the bees will not travel up and over the hazards to take down the syrup. Contact feeders are the answer. Placed over the feed hole the bees will work it readily, especially if the syrup is put on while still warm. The methods used for winter feeding are appropriate, but the mixture is different. For spring or emergency feeding while bees are flying, a weaker syrup is advised. In this case one pint (½l) of water to one pound (½kg) of sugar. To obtain this easily, measure the depth of the inside of the container and divide by five. Make a mark one fifth down from the top. Fill with sugar to this mark (four fifths up to the top), then fill with hot water to the top mark. Stir and feed in the usual way.

If the bees are on their last legs, that is, if there are few live bees, and those which are still alive are falling off the frames, you might need to dribble syrup onto the surface of the comb to keep them going long enough to come up and take the feed for themselves. To be in this situation, however, you must have skimped on the autumn feeding. The colony, if it survives, will take most of the rest of the year to recover — yet another good reason for good autumn colony management.

Back to the first spring inspection
Assuming you find stores on several frames, you can forget the idea of colony starvation for a while. You are now looking for brood. If you find eggs, larvae and sealed brood, have a quick look for pollen in an arc on top of the brood and be satisfied that everything is all right. Close up the hive, leaving the feed hole in the crown board open, and note the date of inspection and brief details of what you have observed. Do not bother to find the queen if you noticed eggs present. It is not worth chilling the brood while you search for her. The whole inspection should take little more time to do than it does to read about.

Spring expansion

In Britain the spring expansion of a colony of bees depends very much on the weather. Assuming the bees have sufficient food reserves to take them through the danger month of March, they will start to bring in nectar, given reasonable weather, from early April, and the danger from starvation will then normally be over.

Perhaps more important from the point of view of a rapid spring expansion is a continuous supply of early spring pollen. Without it the bees cannot rear brood, and without brood at this stage the colony will not expand sufficiently fast to take advantage of the approaching spring blossom. In some areas there is a shortage of pollen at this time of year; however, in many cases the pollen from willow, crocus, snowdrops and many other plants is there but the bees cannot collect it because of adverse weather. Some commercial beekeepers monitor the spring expansion carefully and feed their bees on some form of pollen substitute. Finely ground soya bean flour is accepted and used by the bees for this purpose. However, a colony which has been artificially stimulated to produce extra brood in this way must be carefully husbanded and sometimes fed from then on, since the much expanded colony is more vulnerable to adverse weather later in the season.

From the beekeepers' point of view spring management is a matter of inspections when possible. The bees are better left alone than disturbed in cold, wet, or windy conditions. Only open up the hive when the sun is shining and the bees are flying well. A quick look on a couple of occasions is all that is required before May and the start of the swarming season.

You should satisfy yourself that:
1 the bees have sufficient food and some pollen;
2 worker eggs, larvae and sealed brood are present;
3 the brood nest is steadily expanding across the frames with each inspection, and that the queen still has 'room to lay';
4 the queen is marked to make her easier to find when this is necessary.

If on any occasion you find no food or very little food, turn to the section on emergency feeding (see p. 69).

If there is no pollen, there is likely to be little unsealed brood and probably not much brood at all. Apart from feeding pollen substitute there is little else you can do except keep your fingers crossed that conditions will improve and the colony will catch up on their expansion as quickly as possible.

If you find the queen and she is not marked, mark her there and then (see p. 26).

If you find only patchy drone brood and drones present, suspect a drone-laying queen, or laying workers (see p. 87).

If the mouse guard is still in position, remove it until next winter.

If the brood nest is only expanding at a very slow rate, suspect the disease nosema (see p. 100).

If, on the other hand, you observe your bees alive and well, with brood, stores and pollen steadily spreading across the frames of the brood body, just note their progress in a notebook of some sort and look again a week or two later.

Brood spreading

This is the name given to the system whereby the queen is encouraged to lay eggs at a faster rate than she would normally do were she left to her own devices. By the middle of April a colony which is building up to strength well could have a pattern of brood across the frames of the brood body rather like diagram 1.

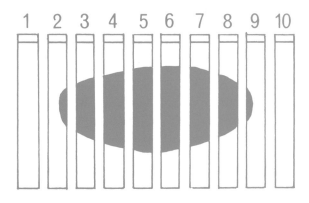

Diagram 1
Brood pattern of well developed colony in spring.

If, during an inspection, the brood is spread, frame number 7 could possibly be swapped with frame number 8, and frame number 3 placed between frames 5 and 6. The brood pattern would then look like diagram 2.

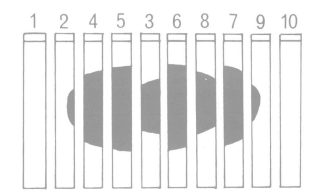

Diagram 2
Brood pattern after brood spreading.

The bees appear to like the brood nest to be a nice neat symmetrical shape. The brood spreading has upset this pattern and the queen, finding areas on frames 2, 3, 8 and 9 which are unoccupied by brood, gets busy and lays up to restore the pattern. Her rate of lay has thus been artificially stimulated.

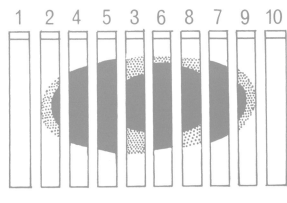

Diagram 3
Brood nest some days after brood spreading. Shaded areas show where egg laying has been stimulated.

The rules of brood spreading state that the frames containing brood can be juggled, but that the brood nest must never be split into two by a broodless frame. This would act as a natural barrier to the queen and one side of the split brood nest might then be neglected so that egg laying would be depressed, not stimulated.

My own opinion is that brood spreading is rarely necessary for the hobby beekeeper. Left to their own devices bees seem to survive satisfactorily, and the beekeeper who artificially stimulated his bees to produce extra brood one month, might have to feed them when their food runs out the next.

When to super

At some point you should find that you have brood on six or seven frames and that the bees, when viewed from above, are thick on eight or nine of the frames, leaving only the odd frame each side of the brood nest uncovered. This is the time to provide extra space by 'supering up' (see p. 42). If conditions are favourable you could remove the crown board a week or two later and find the bees are covering thickly all but the outside frames of the super. This is the cue to place a second super on top of the first. When you have reached this stage the rest is up to the bees. If you are living in a normal, rape-free area, the supers can stay on until they are filled and capped in the autumn. In fact, if you own a really good colony which does not swarm, you might need to add a third and even a fourth super to give your bees storage space to deal with a bumper main summer flow.

You, however, as a beekeeper must be aware of two possible problems which can upset this ideal situation. The first is that your bees may try to swarm, and the second is that you might be in an area where oil seed rape is grown. If the latter is the case, you will have to vary your methods of beekeeping considerably.

Had this book been written a decade or so ago it is most unlikely that the author would have devoted a whole chapter to the subject of oil seed rape. The crop might, however, have been mentioned in a chapter on bee forage. At that time it was rarely encountered in Britain and its relevance to beekeeping was confined to the fact that when worked by the bees, this brassica crop produced honey which granulated rapidly.

Even as recently as the late 1970s, a blaze of yellow was a rare enough sight in the countryside to cause the curious to enquire the name of the crop. Since those days the crop has been more widely grown every year, helped in part by a heavy subsidy paid to the farmer. Beginning as a crop grown mainly in the south of Britain, it has been creeping steadily north every year and shows every sign of becoming more and more popular, as long as insect pests can be kept in check and the profit margin to the farmer remains high.

There are two types of rape, either winter-sown rape or spring-sown rape. Winter-sown rape is much more widely grown at present, since it appears to fit in better with the farmers' rotational plans. Both varieties produce yellow flowers which, when fully in flower, are of such an intensity that to look at a large field can almost hurt the eyes. When pollinated the flowers later produce long, thin green seed pods in which are formed the hard, round rape seeds. These look very much like seeds of mustard, which is botanically a close relative of rape.

Winter-sown rape usually flowers very early in the year. The odd flower in each field can often be seen in mid-April, and by the end of that month the fields are often full of these flowers, remaining so for most of May when they suddenly die off as quickly as they appeared. The farmer leaves the seed to form, then cuts the crop, leaving it in rows to dry for some time before harvesting. When crushed the seeds produce an oil which ends up in a variety of edible products or, less often, as a non-petroleum based lubricant.

From the beekeeper's point of view the crop is good and bad news together. He certainly ignores it at his cost. A field full of rape can yield a phenomenal amount of nectar; nectar, moreover, of a fairly high sugar content which, from the bees' economical point of view is worth collecting. As the crop flowers early in the year when there is often not much other bee forage about, and because the concentration of the crop means the bee can obtain a load of nectar in a comparatively short time with little effort, bees will certainly work it with gusto.

It is fairly easy to tell a bee which is working oil seed rape. She will return to the hive not only with

who find themselves more than three miles from the nearest blaze of yellow in May, can ignore rape; the rest of us have just got to learn to live with it. Our bees will work it whether we want them to or not — and often we would prefer them not to!

A beekeeper in the pre-rape days of the 1950s and '60s watched eagerly for the flowering currant to herald the early spring flow of nectar from the flowers in May, let the honey remain in the supers in case it was needed by the bees in the June gap, and hoped for good weather for the main flow in July. He didn't need to take off any honey until September; in fact, it was considered bad practice to do so. Try this these days if there is rape around, and in September you discover the early rape honey has granulated in the combs. The honey cannot be spun out with an extractor and the granulated super frames cause the beekeeper a major headache.

It cannot be denied that given good weather in May the rape crop yields lots of nectar. Since the 1960s many hedgerows and meadows have disappeared and the rape crop often provides the majority of the bees' annual crop of honey. As previously stated, the problem with rape is that it granulates very quickly, and if the honey gets cold in the comb the speed at which it granulates increases tremendously. Given a few days of nice weather in April or May the bees quickly work the rape, which out-yields anything available at that time in both quality and quantity.

During these months, however, and even into June, the nights can still be very cold. This means the bees from the supers are often needed in the brood body to maintain the temperature of the brood, which has been rapidly increased in response to the stimulus of the sudden nectar flow. Rape honey in super frames not covered with bees at night starts to granulate very quickly, and once started the process speeds up and the frames can go solid in a few days. Therefore, if the beekeeper notices a rape field near his apiary, he should be on the look out for any pollen-covered bees and if he discovers the bees are working rape he must take the honey off before it granulates. In many cases rape honey-filled combs will start to granulate before they are capped over so that one cannot wait for the cappings to appear, which is the normal sign that

her pollen baskets stuffed with golden yellow pollen, but also with a liberal coating of the same pollen over her head and body. Some beekeepers observing this and knowing they were some distance from the nearest rape field must have measured the distance as it seems to be generally accepted that bees will travel up to three miles to work rape, often flying over nearer sources of nectar to reach the crop they appear to prefer.

From this it will be seen that only those beekeepers who live in the middle of a town or city, or those

honey is ready for taking from the hive.

Taking rape honey can sometimes be a problem. The normal clearer board method of removing honey, while probably quite satisfactory for a beekeeper with one or two hives, is considered unsuitable by many beekeepers. As soon as the clearer board is inserted the rape honey gets cold and granulation speeds up. Many beekeepers rely on the tedious and time-consuming method of brushing the bees from the combs, which are transferred to an empty box one by one. A quick shake of a frame held horizontally over the hive will quickly indicate that the honey can be taken. If nectar flies out the frame is replaced: only those which don't shake nectar are taken.

For reasons not yet fully understood, bees working rape, especially those moved on to the crop, can also be expected to be less good tempered than when they are working other crops. Brushing them from combs in this way does not improve their temper, so adequate protective clothing should be worn for the operation. This is definitely not the kind of operation

to carry out on a sunny Saturday afternoon in a built-up area while your neighbours are all out mowing their lawns or taking in the sun.

Having taken the combs from the hive, they should be kept warm and the honey extracted as quickly as possible. Before extracting, it is advisable to shake each frame above a large dish or tray to get rid of any nectar, as rape honey taken from frames still containing some nectar will almost certainly ferment. Fermenting honey bubbles its way out around the lid and spreads itself in a sticky mess down the side of the jar, and the smell is far from appealing. Any shaken-out nectar can be fed back to the bees in a rapid feeder and will not instantly granulate a second time when treated in this way.

Having shaken the frames, they are next uncapped where necessary and the honey extracted in the normal way. Extracted honey needs filtering and the sugary granules in rape honey clog filters in no time at all. Any heat applied during extraction slows the granulation process and some beekeepers use portable electric heaters to raise the temperature of the extraction room to almost tropical levels. If one attempts to filter cold rape honey through a cloth the granules fill the holes in the material and the flow is reduced to a trickle. Honey left to filter overnight is still there the next day, making it a frustrating and tedious business. Here again, heat is of the essence. Probably the best method is to strain the honey straight from the extractor through a coarse metal filter, to remove the thickest of the wax, collect it in a large container and then heat it before filtering. Large-scale beekeepers have special warming cabinets for this purpose.

Unheated honey which has been filtered and bottled immediately tends to go rock hard and form frosted patterns on the inside of the glass. Both these characteristics will reduce your customer appeal, so many beekeepers advocate allowing the rape honey to granulate completely in tins, then reheating for twenty-four hours at 115°F before filtering and bottling. Honey which has been treated in this way appears to stay softer in the jar.

Obviously warming cabinets are not worthwhile for the small hobbyist beekeeper, but heat does seem to be the answer to the problem and Aga-style cookers, airing cupboards and boiler houses can all

be used. Heating for at least twenty-four hours at a temperature never exceeding 120°F seems to produce the results.

Should you find yourself with granulated combs of rape honey you have a worse problem. The sugary granules fill the cells and cannot be spun out with an extractor. If they are given back to the bees and more honey is put in the cells, 'seeding' takes place and the newly added honey also quickly granulates. Even the bees do not seem able to cope well with granulated combs. They need large quantities of water to re-dissolve the granules, but they tend to suck them dry instead, leaving them in the cells where they seed next year's crop. Most beekeepers advocate scraping the frames down to the midribs and heating the resulting messy mixture over a Pratley cappings tray to separate the honey, which again becomes liquid and can be run off from the wax.

Perhaps one of the best plans for oil seed rape honey is to turn as much as possible into wax by providing plenty of foundation at all times. Since more than four pounds (about 2kg) of honey is required to make one pound (½kg) of wax, foundation drawn out from rape honey can save more saleable honey.

Much of this chapter may not be relevant to the system practised by the hobby beekeeper. If he is only interested in honey for his own consumption he can possibly bottle it unfiltered, and if the odd frame goes granulated it is not the end of the world, yet many beekeepers do experience problems with oil seed rape, so this should be seen as a reference section to turn to should the need arise.

Spray damage

Another major set-back of having bees near rape fields is the use of insecticidal sprays. Most oil seed rape is sprayed with pesticide at some time during the year — modern farming is big business and farmers, who stand to gain or lose large sums of money, bring in the big guns when pests attacking a crop reach such numbers that the crop may fail. In most cases the crop is sprayed with insecticide which kills the pests and any other insects either feeding on the crop, feeding on weeds in the crop or merely in transit. Many colonies of honey bees have perished in recent years through this type of spraying. If you are unfortunate enough to find large numbers of dead bees in or just outside the hive, suspect spray damage. Collect three samples, each of about three hundred bees, in paper bags and store them in a freezer. One sample should be sent to the National Beekeeping Unit, Luddington, for analysis (see p. 110). Next, inform your local beekeepers' association. Try to obtain a copy of the BBKA leaflet 'Protecting Honey Bees from Pesticides'. This excellent pamphlet deals with all aspects of the problem, lists all the details required, both for a successful analysis and for a possible legal battle afterwards, and gives advice on starting legal proceedings, advice which hopefully you will never need to seek. Remember, farmers will rarely know about your bees until you tell them, so this should be your first course of action.

Swarming

Without doubt the single biggest reason why many new beekeepers give up beekeeping altogether within a few years is because they fail to understand, and therefore are unable to control, the swarming phenomenon. They know it does happen, but hope it won't happen to them. When it does there is sheer panic, the swarm departs, the depleted colony produces no honey that year and after a year or two of effort and expense and no returns the beekeeper sells up and turns to something else. On the other hand, a beekeeper who understands the swarming process is well on the way to understanding swarm control methods which usually prevent the bees 'doing their own thing', so read this section carefully if you wish to avoid frustration, anxiety and disappointment.

First of all, swarming is a perfectly normal phenomenon. It is a well-known fact that living creatures must multiply to survive — in evolution creatures which do not multiply eventually decline in numbers and become extinct. In the bee world the individual bees do not multiply in the accepted fashion. Their numbers wax and wane according to the seasons; when an old queen is replaced by another there is still only one colony. One colony is vulnerable to disease, starvation, fire or attack by man or beast, and may be destroyed at any time. To overcome this risk of obliteration colonies have developed the ability to split into two so that there are two colonies where only one existed before. In other words colony division, or swarming, has taken place. Of course things can go wrong and one half of

the original colony might not survive but there are now two chances of success instead of one.

How swarming is accomplished is a highly organised process. The bees choose the time of the year carefully. They can only take as much honey as they can carry with them, so they choose the weather and time to suit them best. It is rare to have swarms in Britain before May and as rare to find them after the end of July. May, June and July are the three principal months for nectar production, so that a newly swarmed colony can hopefully collect fresh stores and ward off starvation.

Some time before swarming takes place the queen lays a fertile egg in a queen cup. This is a queen cell in embryo, about eight to ten millimetres across and about the same depth. If the bees decide to use that queen cup (not all queen cups are used), the walls are lengthened and strengthened until the cell is twenty to twenty-five millimetres long and hanging down more or less vertically. The egg hatches out in the normal way and is fed by the nurse bees on a copious supply of royal jelly. After five days as a grub, the cell is sealed by the bees and the queen is left to pupate and emerge a week later.

Before she emerges, however, the old queen will normally have swarmed, taking with her roughly half the bees in the colony. A few days before swarming takes place the workers reduce the food supply to the old queen to slim her down and get her into good flying condition again. Then, on about the day the queen cell is sealed, swarming occurs. Bees often appear to choose to swarm about 10 o'clock in the morning or about 4 o'clock in the afternoon. Here, again, there are good reasons for so doing: a hanging swarm will generate a lot of heat at its centre so that a swarm in the relative cool of the morning or the late afternoon will probably be more comfortable than one in the midday sun. Having said this, you will no doubt find that your bees, if they decide to swarm, will do so just as you are having your lunch!

If the beekeeper were to visit his hive an hour before the emergence of a swarm he could be unaware of what is about to happen. The bees seem to be going about their tasks normally and activity at the entrance is perfectly normal. Inside the hive, however, the behaviour changes shortly before

swarming. The workers which will swarm fill their honey sacs from the combs and wait around in a lethargic manner. Then all of a sudden they pour out of the hive and take to the wing. The air around the hive seems to be full of bees flying around in every direction. The noise is terrific and it appears as if the bees are having a ball. Somewhere in the midst of this swirling mass is the queen, and she and a group of workers will normally settle on some twig, branch or other object before long. Other workers detect her smell and join her, and in no time a swarm forms, composed of a fairly solid mass of bees. This swarm will become the new colony and the bees have to move fast to survive. They must find a home and move into it before their two days' supply of honey is exhausted otherwise they may all perish. The older worker bees, who have been collecting honey, immediately spread out in all directions and inspect every nook and cranny to assess its potential as a new home. If they find a suitable space and decide to move in, they indicate its location by dancing on the backs of the bees on the outside of the swarm. Suddenly the swarm breaks up again and the bees make for the selected home. The leading bees fan their wings to guide the rest and in no time the bees pour into the new home and set to work. The older workers start producing wax again, a comb is formed, and in a day or two the queen will be busily laying eggs once more. A new colony is now in business.

Things can go wrong, however: a new home may not be available, or the weather may change and the bees get drenched, so not all swarms have a happy ending.

Meanwhile let us return to the colony from which the swarm emerged. Half the workers departed with the queen and the half that are left are mainly younger bees. The colony has no queen, but appears happy in the knowledge that one will soon emerge and the bees settle down to business as normal. After a week, on the fifteenth or sixteenth day since being laid, the new queen emerges from her cell, helped by the workers who chew round the cell end. The queen dries off and is immediately groomed and fed by the workers. She spends a few days in the hive, then on a fine day she goes out on her mating flight or flights. She mates with a succession of

Comparative size of a cast swarm

drones before returning to the hive. After a few days she begins to lay eggs and the colony is again a viable unit.

Unfortunately swarming is rarely this simple. The bees usually insure against the chance that they might not raise a new queen by setting out to raise more than one. In most cases there is more than just one queen cell — there can be up to twenty! The presence of this number of queen cells does not affect the old queen's departure with the swarm, but complicates what happens subsequently in the old colony.

When the first virgin queen emerges one of several things can happen. Sometimes the queen will emerge before her sisters, seek them out one by one and sting them to death while they are still in their cells. In this she might be assisted by the workers. If two or more queens emerge more or less simultaneously they will probably engage in mortal combat; one will be killed and the victor will take over the hive. It is possible, however, for the first emerging virgin queen, realising there are more queens to come, to take off with half the remaining bees in the same manner as the old queen. This 'second swarm' is called a cast and stands less chance of survival than a first or 'prime swarm', having fewer worker bees and an unmated queen. Its departure also seriously weakens the original colony, which now has only a quarter of the bees it possessed a week before. It is possible for a colony to literally swarm itself out of existence by throwing out one cast after another, each with its virgin queen and half the available bees, but fortunately this is a rare occurrence. Usually the workers decide to take a hand, accept one virgin and destroy the rival virgins while the colony still has a good chance of survival.

This then is the completely natural swarming

process. The urge to start a new colony is tremendous. The workers of a recently swarmed colony will produce wax and build comb at a faster rate than at any other time. If you recognise the normality of this sequence of events it will help you greatly in coping with the situation when it arises, and it will!

'Swarmy' and 'non-swarmy' bees

Having described the swarming process, it must here be stated that swarming does not happen every year with every colony. Modern research would seem to indicate that the pheromones, or queen substance, which are licked from the queen by the workers and passed round the colony, in some way inhibit the swarming impulse. A young queen exudes more queen substance than an old queen, and a colony with an old queen is more likely to swarm than a colony with a young one. It is said that a colony with a one- or two-year old queen is unlikely to swarm, yet they do so constantly! Some bees appear to swarm annually, some every two or three years and some only very rarely or not at all.

Some colonies with old failing queens raise one queen cell and produce one virgin queen. This virgin goes out, mates, returns to the hive and commences to lay next to the old queen, who has carried on laying without a break. The old queen then usually vanishes after a time and the young queen heads the colony. This process is called 'supersedure' and is comparatively rare. It is also extremely unlikely that a beginner would be sold any supersedure bees, which are normally cherished by their owners!

Many beekeepers draw a distinction between supersedure and 'queen replacement'. When a colony has swarmed, the old queen is sometimes replaced later in the year without fuss, in the same way as that just described for supersedure. Because swarming has taken place, it can be said that the bees have 'replaced' the queen, not 'superseded' her. Most beekeepers, however, use the term supersedure much more loosely than this.

From the beginner's point of view he must expect his bees to swarm every year. If he started with a swarm, they probably will do anyway. If he expects swarming and is prepared for it, he is in a position to

control proceedings. If his bees do not swarm he has lost nothing except his time and the investment in his spare hive. If he hopes for the best and loses a swarm he will have lost any chance of a honey surplus for that year.

Controlling swarming

The swarming process appears to be so deeply ingrained that once it is in motion it is difficult to stop or control, yet this is exactly what the beekeeper has to do if he is to have a chance of obtaining a surplus of honey from his bees. Whereas for the bees it is an advantage to have two separate colonies, albeit weak ones, for the beekeeper this is not such an advantage. The two colonies are so busy rearing brood to reach full strength that they rarely produce a honey surplus that can be removed.

Moreover, the beekeeper has had to provide an extra hive for the second colony. If the beekeeper wants to have a surplus of honey he must try to keep the colony together and thwart the swarming impulse.

It is possible to do this in a variety of ways. None are said to be completely reliable, but probably the method most widely used is the artificial swarm. This is the method I use and is here described in great detail. The beginner is recommended to study and follow it carefully if, and when, necessary — it works for me and should work for you! No other systems of swarm control are described in this book, although there are a number of them. Many are varieties and refinements of the artificial swarm, named in several cases after the beekeeper who first developed the system. To describe other methods could cause complete confusion to a beginner, so it is strongly suggested that you learn to use just one, the artificial swarm method, and stick with it for a year or two before branching out into Snelgrove, Pagden, Demaree, or other methods.

To operate the artificial swarm a spare hive of the same type as that occupied by the bees, or at least one taking the same sized frames, is needed. This last qualification is made because the two most widely used hives in Britain, the National and WBC, both take British National frames. Old WBC hives can sometimes be obtained very cheaply and, although they are not popular with many beekeepers, they make an adequate and cheap second hive. Also needed are ten frames fitted with foundation, or ten drawn combs. It cannot be too strongly stressed that the beekeeper with one hive should be on the look-out for a second hive as soon as his bees are installed in the first. Do not leave it until the bees decide to swarm and then make a forced purchase of hive, frames and foundation. Every beekeeper wants foundation in May and June and sometimes demand outstrips supply, so buy early.

At the beginning of May, assuming the beekeeper has his spare hive purchased, checked over and sitting ready with ten frames of foundation inside it, he should start a weekly inspection of his colony.

Every seven days the hive is gently smoked, the supers and queen excluder removed and the frames in the brood body checked for queen cells. Queen cups will probably be there the first time he looks, but should not cause panic, as not all will be developed to become queen cells. The standard way of inspecting the brood box is to remove one of the outside frames, which will probably not contain brood, and keep it out of the hive near the entrance for the duration of the manipulation.

With one frame out there is a space in the brood body; the frames are inspected, one at a time, and then put back in this space, not where they were before. The space thus moves across the brood body. The frames being inspected should be held vertically at all times. If they are held horizontally nectar will run out of some cells, get the bees excited and the beekeeper very sticky. All the frames are inspected in this way, one by one. If no queen cells with larvae are found, all the frames are returned to their original positions, the frame which was first removed is replaced, and the hive is closed up. A puff or two of smoke every time a frame is removed will be sufficient to subdue the bees and prevent the smoker from going out. More smoke than this can have an adverse effect and make the bees more difficult to control.

If queen cells, either sealed or unsealed, are found, the bees are making swarming preparations and the beekeeper must take steps to prevent this. If all the queen cells are destroyed these preparations can be thwarted. Pushing a hive tool through the cell is usually sufficient. This will set back the swarming preparations for a week and the bees might just give up the idea. To go on breaking down the cells every week should prevent swarming, but sooner or later one cell will be missed and this is all that is necessary for a swarm to emerge. For this reason it is suggested that the beekeeper who finds cells for the second time, or the first time if he wishes to get the business over, moves up his spare hive and proceeds directly to the artificial swarm.

The artificial swarm

Basically, in the artificial swarm the beekeeper takes away the old hive from its site to a position two or three feet to one side, finds and removes the old queen and puts her, and the frame on which she was found, in a new hive on the old site. Within a few

hours all the bees flying from the old hive will return to the old site and enter this new hive. The queen will then find herself in a similar situation to the one she would have been in had she swarmed, namely with lots of older flying bees and a new home, so she and the bees get down to the job of building up the strength of the new colony. The workers produce wax and build comb, the queen starts laying within a day or two and, hopefully, by the time they have built the colony up to full strength the urge to swarm again will have disappeared.

Meanwhile, in the old hive the flying bees have all gone and the workers settle down to wait for the first virgin queen to emerge from the queen cells. If the beekeeper could be sure that only one virgin queen would result, nothing further would be required on his part, but in practice two or more virgins usually emerge, which can cause problems. There is a fairly simple way to ensure that only one virgin survives. It involves putting the bees in a situation where they destroy all but one queen cell and therefore do the job themselves.

The step-by-step stages in carrying out an artificial swarm from the time queen cells are found are as follows:

Assuming that the roof, crown board, super or supers and queen excluder have been removed from the old hive for the inspection, the rest of the old hive, i.e. the floor, brood body and frames of bees and brood, is picked up and placed some two feet to the side of the old hive. The floor and brood body of the new spare hive are next placed exactly on the old site. Then the beekeeper inspects every frame in the old hive carefully, looking for the queen. Even with a marked queen this can be frustrating and time consuming; without a marked queen the beekeeper needs skill, luck, or eyes like a hawk. When she is found the frame she is on is held up for inspection, with care being taken not to knock her off or lose sight of her. She will normally be on one of the middle frames with some brood present. If any queen cells are also present on her frame, either sealed or unsealed, they should be destroyed. This frame, plus the queen, is then placed very carefully in the new hive. Next, a frame of food is removed from the old hive and placed in the new hive. This is to guard against starvation should the weather take a sudden change for the worse.

Finally, the remaining spaces in the brood body of the new hive are filled with frames of foundation or frames of drawn comb. The queen excluder, supers, crown board and roof from the old hive are placed on the new hive on the old site and the operation, as far as that hive is concerned, is complete. Within a few hours all the foraging bees from the old hive will return to the old site and enter the new hive, and swarm conditions will have been artificially created. Apart from a quick inspection a week or so later to reassure the beekeeper that eggs are present and that all is well, this new hive should need no further regular attention for several weeks.

Returning to the old hive, now depleted by a frame of stores, a frame of brood and the old queen, the beekeeper next inspects the remaining frames one by one. There will almost certainly be unsealed queen cells with larvae, and some sealed queen cells could also be present. The beekeeper selects and leaves the two largest unsealed queen cells. One of these should produce the next virgin queen, so they should be treated with care and not knocked, bumped or inverted. Next, all the other queen cells, both sealed and unsealed, are destroyed either by removing them completely or by crushing them with a hive tool. The frames are replaced and a dummy board used to block off the gap left by the two which are missing. While inspecting the combs, the beekeeper should also take note of the amount of stores in the old hive. Unless the frames are packed with plenty of honey this colony could starve in the next few weeks. The beekeeper has just artificially robbed the colony of its older foraging bees and so almost no nectar will be brought in from outside by the bees. It is therefore a good idea to feed the colony at this time. To do this the beekeeper places the new crown board on the old hive and adds a spacer, or eke, before replacing the roof. Then at night he can return with a feeder of sugar syrup, invert it over the feed hole and let the bees do the rest. They will not take the food if they don't need it so nothing is lost by this feeding.

With few old bees to defend the stores against bees from other colonies there is a strong possibility that robbing will take place if the colony is fed sugar syrup. To minimise the risk of this a reduced

entrance block should be fitted.

A week later (in about seven or eight days) the beekeeper performs the second much simpler operation and then the whole business is complete for the year. The old hive is picked up bodily and carried from one side of the new hive to a convenient position three feet or more on the opposite side. This operation is somewhat awkward and requires two people. The timing is critical and the move *must* be made on the seventh or eighth day, even if it is raining or blowing a gale. In fact, a wet day makes it more easy to persuade a non-beekeeper to help you since the bees will not be flying. The move can also be made in the late evening for the same reason, but watch your footwork when stumbling around in the dark with hives of bees!

The reason for this second move is to remove more foraging bees. Any bees which have become foragers in the seven days since the first move will have become accustomed to flying from and returning to the old hive on the new site. When the second move is made the bees flying from the old hive will return to find it gone. They will search around for some time and eventually find their way into the nearest hive, which will be the new hive on the original site. The result will be that the old colony will again be short of foragers and the workers themselves will sense this and destroy all but one of the queen cells. A few days after the second move the remaining virgin queen should emerge. Some days later she will leave the hive on brief flights, subsequently mate with several drones, return to the hive and shortly after come into lay.

During this period the beekeeper should stay away from the hive as his presence could upset the natural chain of events. The recommended course of action is to stay away from this colony for three weeks after the hive is moved the second time. If after three weeks the beekeeper inspects the colony and finds a small patch of eggs, or even eggs and larvae, everything is fine and the colony can be left safely for some time with no further action required. If no eggs or brood are seen the beekeeper should go away and look again in a week's time. If there are still no eggs by this time something has probably gone wrong. The queen could have failed to emerge, failed to mate successfully, or failed to return to the hive after mating. In this case, the queen-less colony cannot survive, so it is probably best to leave it another week before checking it again and if it is then still queen-less, unite it to the colony containing the old queen.

It is strongly recommended that the two colonies be reunited in any case. If all went well the bees with the old queen in the new hive will have got on with the job of collecting honey, at the same time expanding up to a full colony. The young queen should also be laying fast and expanding her brood nest. All this will have taken four or five weeks from the time queen cells were found, which is about a third of the nectar-collecting season. If the old queen is found and removed and the two colonies are united, the resulting strong colony (with a young

laying queen) will produce more main crop honey, require only one hive to be fed for winter and, more importantly, leave you with your spare hive again for next year's swarm control.

Uniting

Uniting is the name given to the operation where two separate colonies are joined together to make one stronger colony. A weak colony consumes most of the nectar its bees gather to maintain the brood temperature, rarely produces a honey surplus and is much more likely to die out during the winter than a strong colony. By uniting two such weak colonies the beekeeper increases the honey-gathering potential and the chances of survival. There are other good reasons for uniting and more than one way of doing it, but uniting by the newspaper method is the safest and easiest.

Both colonies are inspected. The queen which is to head the united colony is checked to make sure she is laying well. If worker brood is observed the colony is said to be 'queen-right'. The other colony is then inspected and the queen is found and removed. A sheet of newspaper is pinned over the top of the brood body of the queen-less colony and the brood body of the queen-right colony is taken from its floor and placed on top of the newspaper. The crown board and roof are replaced on top of this double brood body and the first manipulation is complete. The bees in the top box find themselves imprisoned and start to chew their way through the newspaper. Some beekeepers stab a few small holes in the newspaper to ease the bees' task, but this is not strictly necessary. By the time the two lots of bees have chewed through the newspaper the different smells of the colonies will have intermingled: the queen-less colony will have realised this fact and be keen to rectify their deficiency, and the two colonies should unite without fighting.

After two or three days the hive is opened up and the frames are reorganised. The queen and all the frames containing brood are placed together in the bottom brood body. Frames of food are put on either side of the united brood nest to fill this brood body and the second brood body, and any extra frames are taken away for storage.

Although the queen-right colony should be placed on top of the queen-less colony, this does not appear to be essential as the method seems to work just as well when the queen-less colony is placed on top.

Failing queens and absent queens

The survival of a colony of bees depends to a great extent on the queen, the egg layer. In the summer months thousands of bees perish daily after a short, yet apparently satisfying, life literally working themselves to death. Their places must be taken by other bees so a constant supply of fertile eggs is required of the queen. Several things can, however, happen to upset the continuity of egg supply. Firstly, the queen may die or be killed; the colony is then said to be queen-less. Should this happen when drones are around, that is, from April to September, the workers will set about raising another queen.

They do this by constructing emergency queen cells round young worker larvae of less than three days old. They swamp the larvae with royal jelly until they alter their position to occupy the vertical cell extention, then in five or six days time they seal in the fully grown larvae and in only fifteen days after the eggs were laid a new queen, or several new queens, should emerge. The first queen might kill her less developed sisters and potential rivals while they are still in their queen cells, or she might decamp with half of the bees in a swarm and leave her sisters to sort out the succession. In either case, the virgin queen must be mated within fourteen days or she will go stale. She may start to lay, but will lay only infertile eggs which will all hatch out into drones, even if laid in worker cells. She is then known as a drone-laying queen and unless she is replaced the colony is doomed.

During the mating flights the virgin queen mates

with several drones. The spermatazoa from each is stored in the queen's spermatheca until required for use. It can happen that, because of incomplete mating, the supply of spermatazoa runs out before the life of the queen. If this is a slow failing the bees will sense it and raise another queen, but should this happen suddenly at a time when the bees have no fertile eggs or young worker larvae available, or should the supply run out in the winter months, then raising a replacement queen is impossible. This also results in the queen becoming a drone layer.

Should the queen die over the winter months and the workers are left for several weeks with no queen and no way of raising one, the ovaries of some of the younger workers begin to develop and they start laying eggs. These can only be infertile drone eggs, even if laid in worker cells. These laying workers are prone to lay several eggs in each cell in a random and haphazard fashion on the combs.

Should you suspect that any of the above misfortunes have befallen your colony you should get a second opinion by consulting a beekeeping friend. If, on any inspections, you find a mass of untidy drone brood, lots of drones, few workers and no worker brood, it is quite possible you have a drone-laying queen or laying workers. I strongly suggest you get some help in sorting the problem out.

Dealing with a drone-laying queen

The offending queen must be found and removed and another laying queen introduced. This

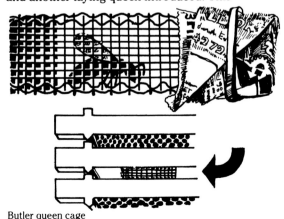

Butler queen cage

presupposes you have another queen available or can buy one, which is not always an easy thing to do. If you are successful in finding a replacement, the procedure is to inspect the frames one by one until the drone-laying queen is found and then to remove and destroy her. Leave the colony for a few hours to realise they are queen-less and then introduce the new queen. This can be done by simply releasing her in the entrance and letting her run in herself, but a safer method is by opening up the hive down to the brood body, smoking the top bars of the frames vigorously and sliding the queen in unnoticed in the ensuing confusion. Another way is to put the queen in a Butler queen introduction cage into the gap between the top bars of two adjacent brood frames and to replace the crown board and roof. The introduction cage is plugged at one end by sugar candy, or a small piece of newspaper held in position with an elastic band, and in time the bees chew through this, thereby releasing the queen. By the time they release her, her smell will have been accepted and so, too, will she.

Dealing with a colony containing laying workers

If a queen is introduced into a colony which has been queen-less long enough for laying workers to appear, she will not be accepted and will be killed. The suggested way to deal with this is to unite the offending colony to one which is queen-right. Not an easy thing for you to do if it happens to your one and only colony of bees, but the procedure is as follows.

The hive containing the laying workers is removed from its site and the queen-right colony is put in its place. Remember the 'three-feet or three-miles' rule. If you are moving one of your own queen-right colonies, you must move it in short 'three feet a day' stages, otherwise the flying bees might fail to find the new hive location and will perish. Then the rogue colony is taken about a quarter of a mile away and all the bees brushed off the frames onto the ground. The laying workers cannot be identified from the other workers by inspection, but they are different in that they have lost the ability to fly. When they are all dumped unceremoniously on the ground the majority of the workers wing their way back to their site, enter the new hive containing the queen-right colony and unite without any problem. The laying workers are left to walk home. Should any of the more determined ones make it to the hive they will not be allowed in anyway, so this is, in effect, a way of sorting them out and destroying them.

Test for a queen-less colony

It is possible for a queen, while quite healthy, to go completely out of lay in the autumn. This is a common occurrence, but most beekeepers would like to be assured that the colony will go into winter with a good queen. There is a simple test which can be used at certain times when queen-lessness is suspected. The test requires a brood frame containing some eggs and young worker brood, which is not always available, especially to the one-colony beekeeper. The frame is brushed clean of bees, placed in the suspected queen-less colony and left for four or five days. A queen-right colony will feed the worker brood in the normal way, but a

queen-less colony will attempt to raise a queen or queens from the eggs or young worker brood. If, after four or five days, you see emergency queen cells your colony is indeed queen-less.

Here, more than at any other time, you should seek the assistance of your beekeeping friend. He will possibly be able to provide the frame of eggs and young brood required for the test and his help will be invaluable in sorting out any real egg-laying problem.

So you blew it and they swarmed!

All experienced beekeepers have lost swarms at some time or another. Some will even admit it! So, if one day you are confronted by a wheeling, diving, darting, throbbing mass of bees, making a noise like a demented vacuum cleaner and forming a cluster in your neighbour's apple tree, do not feel too guilty.

First, cool the swarm down. Water with a watering can or hose pipe but don't wash them all away with a

Calming a swarm with a fine spray of water

direct jet from a hose. Wetting the swarm means you have more time to act, since they are then less likely to move off elsewhere in a hurry. Next, take the roof off your hive and remove the crown board. If there are few bees meeting your gaze it is a fair assumption that it is your colony which has swarmed. This is an important precaution as they may not be your bees but may have come from another colony, either wild or belonging to a neighbouring beekeeper. It is useful to remember this in case the bees are causing panic or annoyance to the public and you think it advantageous to shift the blame. To say, 'I don't think they are mine, but I had better sort them out!' will gain you house points, not lose you them.

Having cooled the swarm, you must plan the campaign of taking it. For this you need an open topped receptacle of some sort. A straw skep is ideal as it is strong and light and can be balanced on one

hand, which is often an advantage. It is also the traditional way swarms are collected and adds an authentic touch of interest to the show you are putting on for the spectators. A light wooden box somewhat smaller than a brood body of a hive is ideal, but a medium-sized cardboard box will do at a pinch. Ideally, the swarm will settle on the branch of an apple tree some two metres high. The box is held underneath the branch which is given a sharp bang, the bees fall off the branch and into the box. The bees cling together and to the side of the box so that when it is turned upside down, which is the next move, the bees do not fall out again. The box is set on the floor beneath the apple tree, with one corner raised by a pebble, stick or brick so that bees can enter or leave. In no time the bees sense their queen is with them and this is now their home. They start to fan, and flying bees return to the box. In the cool of the evening they are all in the box and bees and box can be removed for hiving.

To deal with a swarm so neatly is a rare occurrence. If most of the bees are caught in the box but the queen is still outside they will vacate the box as quickly as they vacated their hive and rejoin her. In this case you try again and hope for better luck. The bees can also make life more interesting for the beekeeper by settling at the top of a chimney or nearby tree. Here you should think carefully about whether it is worth the risk of collecting them. Holding a box or skep with one hand and using the other hand to dislodge the bees does not allow for much of a hold on a ladder.

Bees which swarm into a thick bush or hedge can sometimes be cut out a twig at a time and shaken into the box, but some swarms have to be written off. Swarms which enter cavity walls or nooks and crannies inside houses usually have to be destroyed. Sometimes the box can be inverted over a swarm and the bees encouraged to walk up and in by puffs of smoke from the smoker.

It can be a time-consuming business; the fortunate thing is that the bees for some time after they swarm will probably be in high spirits and not too aggressive. Spectators should, however, be warned to stay at least ten metres from your operations. One might think this is unnecessary, but some children show a complete lack of fear and come close in order to miss nothing.

Hiving a swarm from your own hive

If a swarm emerged from your hive, this almost certainly means that there are queen cells present. If you merely put the swarm back, the chances are they will swarm again next day. You must use your spare hive to re-house the swarm and old queen, and manipulate the brood in the old hive in such a way that the bees allow only one virgin queen to emerge. Apart from the fact that the bees have actually swarmed, the initial and subsequent operations are similar to those previously detailed for the artificial swarm. Hiving a swarm should be done on the evening of the day the swarm emerged as follows.

The queen and swarm are placed in a new hive on their original site. The old hive is moved to one side and a new, empty hive is placed on the old site. A

Bees working extra-floral nectaries on back of laurel leaf

frame of emerging brood, plus adhering bees, is taken from the old hive and placed in the new one. Any queen cells on this frame must be destroyed during the transfer. The emerging brood will usually encourage the queen to stay in the new hive. A frame of food from the old colony is placed in the new hive and the brood body is then filled with frames of foundation or drawn comb, or a mixture of both. An empty eke, super or brood body is placed on top to prevent the bees overflowing down the sides, and the swarm is shaken into the hive. The bees should quickly make their way down between the frames, the spare box can be removed and the queen excluder, supers and crown board can then be placed on the new hive. The roof is replaced and the swarm should go to work to build a new brood nest.

The old hive is next examined. Two large unsealed queen cells are left — all other queen cells, sealed or unsealed, are destroyed. The frames are moved to one side and a dummy board is used to block off the empty space. If there is plenty of food on the remaining frames, simply replace the crown board and roof, and this manipulation is now complete. If there is little food present, put on an eke or feeder, and feed the colony that night. In any case, reduce the size of the entrance with a reduced entrance block.

One week later, move the old hive bodily from one side of the new hive to the other. Bees which have graduated to flying duties during the week will return, find their hive gone and enter the nearest hive, the new one. In this way the numbers of foragers for the new hive are boosted and the depleted number of bees in the old hive causes them to destroy all but one virgin queen. This virgin should develop, mate and come into lay two or three weeks later. Do not examine the hive during this period. As soon as the new queen is seen to be laying well the old queen can be removed from the new hive and the colony united with the old colony containing the new young queen.

Hiving a swarm which is not from your own hive

This is a great temptation, but you are advised to avoid it if possible. Swarms can bring in disease, they can become queen-less or bring in second-class queens, they rarely produce a surplus of honey in their first year, they usually need feeding and, most importantly, they occupy your spare hive and equipment. The day after you use your spare hive for a stray swarm you are certain to need the hive to prevent your own bees swarming. If, however, you decide not to take the swarm, try to ensure that some other experienced beekeeper does, as 'wild' swarms can cause tremendous problems when left to their own devices.

Quick reference pages

The beekeeper's year

January

The queen, surrounded by thousands of her workers, is in a rugby-football shaped cluster in the hive. There is little activity except on a warm day when workers take the opportunity for making defaecation flights. There will be no drones present, but some worker brood will be raised.

Little work is required at this time. If the ground is covered with snow, shield the entrance to cut out the light and prevent workers flying and perishing on the snow.

Estimate less than one hour for the month.

February

The queen, still surrounded by workers in the cluster, lays a few more eggs each day. It is still 'females only' in the hive, and workers again take cleansing flights on warm sunny days.

Once again, you need do little, other than prepare equipment which will be needed in May.

Possibly one hour for the month.

May

With good weather, nectar and pollen can come in thick and fast. The queen will be reaching her greatest rate of lay and there should be brood across most of the brood box.

Add supers as necessary. Some honey can be removed; it must be removed if oil seed rape is grown nearby. Watch out for swarming preparations. Inspect the hive weekly. Have a spare hive ready, and artificially swarm where necessary.

Five or six hours for the month.

June

Unswarmed colonies will be very populous. The queen's rate of lay should drop.

If rape was worked by the bees the honey in the hive will need extracting, in which case care should be taken to ensure that the bees do not starve during the 'June gap'. Keep up weekly swarm control inspections for unswarmed colonies; artificially swarm when necessary.

Five or six hours for the month.

September

Drones are likely to disappear overnight. Hive population is much reduced. Queens often stop laying completely.

Remove the honey. Leave bees to their own devices. Start winter feeding towards the end of the month.

Two or three hours for the month.

October

Very little activity — the bees are preparing for winter.

Finish winter feeding. Put on mouse guards. Check hive is secure for winter.

One or two hours for the month.

What the bees **do month by month, what you,** **the beekeeper, need do to keep them happy and** **how long it should take you.**

In the interests of simplicity and brevity these notes are of necessity generalisations. Weather, climate, neighbourhood and the type of bees will influence the pattern. No two colonies will have exactly the same pattern of behaviour — but then, nor will any two beekeepers.

March	April
This is the month when colonies can die of starvation. If the bees were adequately supplied in the autumn this should not happen, but it still does. With lengthening days the queen steadily increases her rate of lay; more brood means more food consumed and the bees are not bringing any nectar in from outside.	Hopefully the weather improves and some early blossom appears. Flowering currant is usually the earliest nectar yielder, but oil seed rape crops flower at about the same time.
On a fine day, when the bees are flying, have a quick peep inside, without disturbing the bees too much by removing the frames. If no sealed stores are seen, a small quantity of syrup should be fed.	The odd drone should appear and the rate of brood rearing should increase dramatically to give a rapid expansion of the colony. On a fine day when the bees are flying find and mark the queen so that she can be recognised more easily later on. Put on a queen excluder and super of drawn combs when necessary.
One or two hours during the month.	Two or three hours for the month.

July	August
If the weather is good the main nectar flow will occur. The hive population diminishes as the queen's rate of lay drops. Drones are still present.	The colony strength diminishes fast. Drones are still around. Outside activity is reduced since there is little nectar available.
Add supers as necessary. Continue weekly swarm control inspections if still necessary. Rub hands in anticipation if a good nectar flow occurs. Possibly re-unite colonies.	Swarm control inspections are no longer necessary. Possibly restrict the size of the entrance to prevent robbing by wasps or other bees. Replace the old queen with a new laying queen if one has been reared. Forget the bees — you can go on holiday!
Four to six hours for the month.	One or two hours for the month.

November	December
Even less activity. Bees will probably go into a cluster.	The bees are in a cluster.
Store equipment away for the winter.	Enjoy Christmas!
One hour.	Nil.

Collecting a prime swarm in a skep

Summary of pages 88 to 91:

Dealing with a swarm

Don't panic — it's not the end of the world.
Find the swarm.
Warn onlookers to stand well clear, or stay indoors, especially children who often show no fear and approach too close for comfort.
Don protective clothing.
Try to cool the swarm down with watering can or hose pipe.
Swarms often come in from other colonies so inspect your own bees.
Remove roof and crown board.
If few bees are visible, it's probably your own swarm.
Next, 'take' the swarm; a skep is best but a cardboard box will serve.
Hold the receptacle below the swarm, if possible, and dislodge the swarm by sharp movement.
Upturn the box and position on the ground as near to the swarm location as possible.
Bees must be able to enter or leave so prop one side with a stone or stick.
Where the swarm cannot be easily dislodged use ingenuity and natural resourcefulness somehow to get the bees into the box or skep.
Sometimes bees can be driven into the box by gentle puffs of smoke — especially since smoke encourages them to climb upwards (this is a natural tendency).
Once in the box there is no hurry. Take your spare hive and put the swarm (containing the old queen) in it on the old site.
To do this:
First, move the old hive to one side.
Put the spare hive on the old site.
Next, wait until evening.
Then, gently smoke the old colony.
Remove the roof, crown board, supers, and queen excluder.
Select one frame of mainly stores from the old hive and place into the new hive with any bees adhering to it.
Next, inspect remaining frames and select one with no queen cells but some sealed brood, preferably with some emerging brood. Place this, plus adhering bees, in the new hive.
Fill the remaining space in the new hive, ideally with three or four frames of foundation interspersed with

frames of drawn comb, although either, or a mix of both, will suffice.
Put an empty super on top of the brood body to stop bees spilling over during hiving.
Pick up the box (upside down).
All the bees should be clinging together at the top, and the box can be carried carefully like this.
If in doubt place the box carefully on a spread-out sheet of material and then knot the corners before moving.
Hold the bees in the box over the open hive on the old site and bang the top of the box.
Bees should fall onto the top bars and quickly move down onto the frames.
Remove the empty super, if used; put on the queen excluder and supers and close up the hive.
You are now in the same position as if you had just performed an artificial swarm.
Subsequent manipulations are the same.
The next stage is to inspect the old hive, leave two large unsealed queen cells, destroy all the rest (both sealed and unsealed), close up remaining frames with dummy board and close up the hive.
Flying bees from the old hive will return to the swarm in the new hive.
No nectar will be brought in so that the old hive might need feeding.
A week later (7 or 8 days) move the old hive from one side of the new hive to the other.
The old colony should then raise a new laying queen and once this is observed (some 4 or 5 weeks after swarming) the old queen can be removed and the colony united with that containing the young laying queen.

Summary of pages 82 to 85:

The artificial swarm

Pre-planning: obtain spare hive and equipment in good time.
You will need: spare floor, entrance block, brood body, crown board, ten frames with foundation or ten drawn combs, dummy board, feeder and spacer or eke.

Early in the year (March or April): find and mark the queen.
During May, June and July: have spare hive and equipment ready and waiting. Inspect every seven days for queen cells. When found, proceed with artificial swarm.

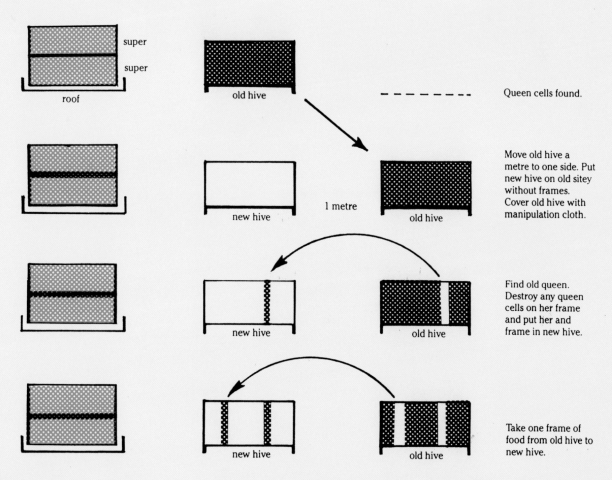

super
super
roof

old hive

Queen cells found.

new hive   1 metre   old hive

Move old hive a metre to one side. Put new hive on old sitey without frames. Cover old hive with manipulation cloth.

new hive   old hive

Find old queen. Destroy any queen cells on her frame and put her and frame in new hive.

new hive   old hive

Take one frame of food from old hive to new hive.

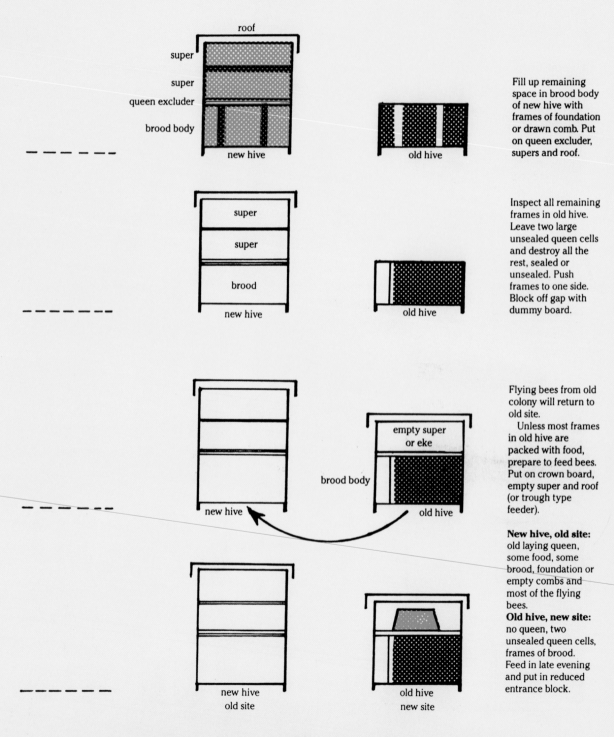

Fill up remaining space in brood body of new hive with frames of foundation or drawn comb. Put on queen excluder, supers and roof.

Inspect all remaining frames in old hive. Leave two large unsealed queen cells and destroy all the rest, sealed or unsealed. Push frames to one side. Block off gap with dummy board.

Flying bees from old colony will return to old site.

Unless most frames in old hive are packed with food, prepare to feed bees. Put on crown board, empty super and roof (or trough type feeder).

New hive, old site: old laying queen, some food, some brood, foundation or empty combs and most of the flying bees.

Old hive, new site: no queen, two unsealed queen cells, frames of brood. Feed in late evening and put in reduced entrance block.

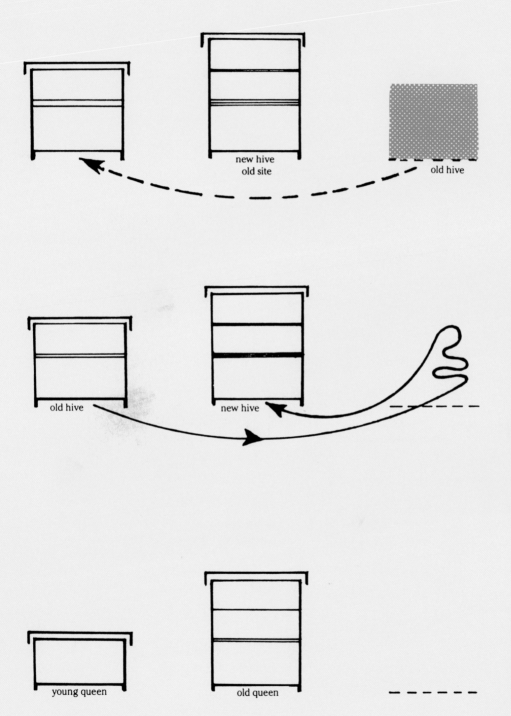

new hive
old site

old hive

old hive

new hive

young queen

old queen

Seven or eight days later: move old hive from one side of new hive to the other.

What happens then is: young flying bees from old hive will return to their site, and finding no hive there, they search around and eventually enter the nearest hive.

The old hive will thus again be depleted of flying bees shortly before the first virgin queen emerges. Only one virgin will be allowed to survive. She will mate and return to lay and head the colony.

Four or five weeks after the artificial swarm the position should be: in old hive a young queen in lay with steadily expanding brood nest; in new hive an old queen expanding brood nest, lots of foragers, a surplus of honey in supers.

This is the time to remove the old queen and unite the two colonies. This will give a large colony and, hopefully, will result in a larger honey crop. Equally important, it will release the spare hive for use again next year.

Uniting

Preliminary actions: check both colonies. Decide which queen is to head the united colony. Ensure she is in lay.

You need: a sheet of newspaper and four drawing pins.

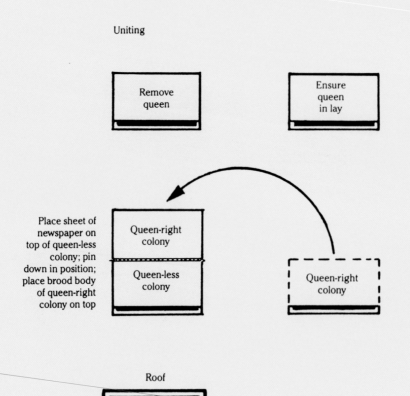

Uniting

Remove queen

Ensure queen in lay

Place sheet of newspaper on top of queen-less colony; pin down in position; place brood body of queen-right colony on top

Queen-right colony

Queen-less colony

Queen-right colony

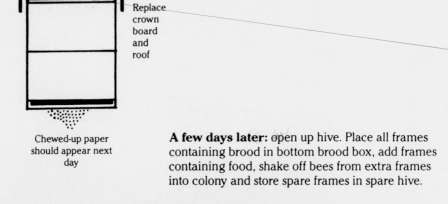

Roof

Replace crown board and roof

Chewed-up paper should appear next day

A few days later: open up hive. Place all frames containing brood in bottom brood box, add frames containing food, shake off bees from extra frames into colony and store spare frames in spare hive.

Entire books have been written on this subject alone, so in no way is this section a full and complete list of the diseases which can afflict bees, or of their symptoms, treatment and cure.

Most of the important bee diseases are covered in an excellent series of leaflets published by the Ministry of Agriculture, Fisheries and Food. I cannot improve on the text of these leaflets and do not intend to try to reproduce them here. To do so would require so much space that it would, I consider, give the beginner a false idea of the likelihood of disease striking down his bees. It is rather like ploughing your way through a medical dictionary — at the end you wonder how you could possibly still be alive.

Generally speaking, a novice beekeeper who has obtained his bees from a reputable source is unlikely to have more than minor ailments affect his bees for years to come, and some of these might even pass unnoticed!

Most experts seem to agree that good apiary management helps to prevent the spread of diseases.

Points to consider are:
1 Take care to buy your bees from a reputable beekeeper.
2 Avoid buying old combs or second-hand frames unless they, too, come from a reputable beekeeper.
3 Scrape down, wash and clean up second-hand hives and, if possible, also scorch wood inside with a blow-lamp before using in your apiary.
4 Do not leave broken frames or odd pieces of brace comb lying about. Preferably recover wax from these sources, but if this is not possible dispose of them.
5 Do not feed honey from doubtful or unknown sources back to bees or let them clear out the remains of honey jars because honey may contain spores of brood diseases.
6 Always try to prevent robbing, as robbing bees may take diseases back to their own hives.
7 Keep your ears open for the presence in your area of any wild or neglected colonies of bees. They, too, could be a source of disease.
8 Try not to move your bees to the crops too often in any one year. One disease, nosema, seems to be brought on by stress caused by frequent moves.
9 Be wary about taking or accepting swarms. The bees can bring in diseases to your apiary.
10 If your bees should die out in the winter or spring, close up the hive entrance to prevent robbing. The bees might have died of starvation, but they could also have had some disease which would be spread by robbing bees.
11 Replace brood comb every three or four years.
12 If you have any doubts about any of your bees call in the Bees Officer and take his advice. He is the expert!
13 Read all the detailed MAFF leaflets on bee diseases and keep them handy for reference.
14 Invite the MAFF Bees Officer to conduct regular inspections of your bees. He will be a source of useful advice of all sorts while he is doing so.

The dreaded Foul Brood!

It sounds ironic, but the two most serious diseases afflicting bees in Great Britain are American Foul Brood and European Foul Brood, often referred to as AFB and EFB. With both of these, brood is killed in large quantities and the disease spreads with extreme rapidity.

American Foul Brood

The brood dies in the cell after being sealed in. The signs are sunken and perforated cappings. If you stick a matchstick into an infected cell the diseased larva pulls out in a sticky thread (the ropiness test).

Action: if you even suspect AFB call the MAFF or the Bees Officer.

European Foul Brood

This time the larvae die before being sealed in, adopting abnormal twisted positions. The dead larvae are yellow, turning brown after a time.

Action: contact the MAFF Bees Officer.

As soon as the MAFF is notified of a possible outbreak of disease, the Bees Officer will arrange to inspect the colony and, where necessary, any doubtful combs will be sent for laboratory analysis, since EFB can only be positively diagnosed by microscopic examination. Upon confirmation of disease, the MAFF pursues a policy of killing the bees, by burning in the case of AFB and destruction, or carefully supervised treatment of colonies, with EFB. This has had the effect of limiting the spread of the diseases so that they are not endemic in Great Britain, as is the case in some countries.

Apart from these two it is unlikely that any of the other diseases listed below will reach the stage of decimating your bees.

Acarine

This is caused by a parasitic mite, which lives in the breathing tubes of adult bees and feeds on the bees' blood. A lightly infested colony might show no signs of the infestation, but bees badly infected with the mite are unable to fly and crawl around the hive entrance with their wings at odd angles.

Action: send a sample matchboxful of bees for analysis, either to your local Bees Officer or to the ADAS National Beekeeping Unit at Luddington (see p. 108 for address). If Acarine is confirmed, then follow the treatment outlined in the Acarine leaflet.

Nosema

This very common disease is caused by a microscopic organism, which lives in the gut of the adult honey bee. The signs of the disease can be the fouling of combs and the hive front with chocolate coloured droppings, but usually the only signs are the failure of the colony to build up in strength when it should be doing so.

Action: if nosema is suspected, send a sample of bees for laboratory analysis and, if the disease is confirmed, treat as per MAFF leaflet.

Amoeba

A different bug, but one which produces similar effects to Nosema.

Action: as for Nosema.

Dysentery

Bees will not normally soil their comb. When they do, they are said to have dysentery. This can be caused by a bad attack of Nosema, by the bees eating fermented stores, or by prolonged bad weather which prohibits cleansing flights.

Action: check for Nosema, read the MAFF leaflet on Dysentery in bees and pray for better weather which will possibly put matters to rights.

Sacbrood

A virus disease, which causes the larvae to die after being sealed in, forming a loose skinned sac as they do so. The dead larva dries out and forms a scale, with the head end curled upwards giving a Chinese slipper effect. This can often be seen through holes chewed by the bees in the cappings. Sacbrood is rarely serious and usually disappears towards the end of the season.

Action: if only a few cells — none. If more widespread, re-queening might help.

Chalk Brood

This appears to be relatively common in some parts of Britain. It is caused by a fungus, the spores of which become mixed with the larval food, are eaten

Proper attire for examining bees

by the larva, germinate inside it and kill it after the larva has been capped. Other bees later chew off the cappings to reveal a chalky-white mummified larva which can be picked out from the cell cleanly with a pin. Sometimes the chalky-white appearance turns grey or black as the fungus produces spores to complete the cycle and re-infect other larvae. Usually only odd cells are affected; the disease is commoner in May and June than later in the summer season.

Action: if only a few cells are affected — do nothing. If attack is more serious and does not readily clear up, seek further advice or consult MAFF Brood diseases leaflet.

There are several other diseases or disorders which might possibly afflict your bees and might cause you to panic. The first action should always be to seek a second opinion, an experienced beekeeping friend in the first instance, and then the MAFF Bees Officer if it is deemed necessary. Only by having the differences pointed out to you will you appreciate which problems are serious and which are not.

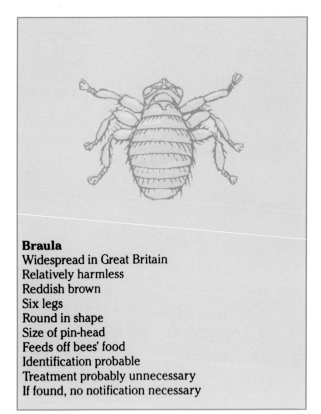

Big bees have little B's upon their backs to bite them!

Two bee parasites — BRAULA and VARROA

Braula

If you look carefully at a frame of bees, you may notice a darkish brown creature, about the size of a pin-head, hitching a lift around the hive on the backs of the worker bees. The wingless fly, BRAULA COECA, or braula for short, scurries from one bee to the next with great agility, and lives off the bees by nipping forward and taking his share during the transfer of food from one bee to the next. Braula have six legs and are roughly circular in shape. They are found in reasonable numbers, a few hundred or so, in some colonies and areas, but in some districts they are a comparative rarity. They normally appear to co-exist with the bees without causing any real problem, but they seem to prefer a royal host and in some cases there can be six to eight all living on the queen, which must cause her some annoyance. They usually travel on the upper surface of the thorax and in this way they can spread from one hive to the next by clinging to a bee which might subsequently drift into the wrong hive. They breed by laying eggs in the cell cappings, the larvae hatch out and tunnel along inside the cappings, pupate and as adults latch onto the nearest available bee.

Under normal circumstances beekeepers with colonies where some braula are present ignore them — as the bees themselves appear to do. It is said, however, that smoke from tobacco temporarily

Braula
Widespread in Great Britain
Relatively harmless
Reddish brown
Six legs
Round in shape
Size of pin-head
Feeds off bees' food
Identification probable
Treatment probably unnecessary
If found, no notification necessary

knocks out the braula, while leaving the bees unharmed, and one course of treatment is to slide a paper beneath the combs, smoke the bees heavily using tobacco, allow time for the braula to drop onto the paper and then withdraw the paper and burn it before the braula come to.

It appears that in any colony the numbers of braula present wax and wane, and I have heard it claimed that the addition of a few drops of creosote in the smoker fuel controls the spread of braula quite effectively.

The main reason that any beekeeper should get to know and recognise these uninvited guests in the hive is so that he can distinguish between braula and its much more lethal look-alike, the varroa mite.

Varroa

The Varroa mite, VARROA JAKOBSONI, or varroa for short, which was unknown to most British Beekeepers until the late 1970s, has regrettably forced itself to the attention of British beekeepers to the extent that beekeepers who ignore varroa completely will probably not have live bees to keep for very long.

The mite, which lived in harmony with its host, the Eastern honeybee (Apis cerana), in

Braula on the bee's thorax

Varroa
Spreading throughout Great Britain
Can kill colony
Reddish brown
Eight legs
Crab-like in shape
1 mm long, 1.5 mm wide
Sucks blood from bees or larvae
Identification difficult in early stages
When present treatment essential
Notifiable pest but presumed present in many areas

the Far East, transferred its attention to European honey bees when they were moved into that area. Other beekeepers moving bees from the Far East, inadvertently spread the varroa mites and with them the disease Varroois, the name given to the honey bee disease caused by the mites. At the time of writing, varroa has been found in England, Scotland and Wales. At this time no cases of varroa have yet been found in Ireland.

The European honey bee, Apis mellifera, has proved much less able to cope with the parasite than her far eastern cousins, and in many cases the mites reach such a degree of infestation that whole colonies perish. From the time that varroa arrives to the time the colony fails could be several years. During this time the beekeeper could have moved his bees to many different sites, they could have been sold or swarmed, or the varroa mites could have spread to many other colonies with drifting bees.

Looked at from above, the adult female varroa are crab-like in shape, about 1.5 mm wide and only 1mm from front to back. They have eight legs, and the adults are reddish brown in colour. The adult females hitch a lift around on the bees, usually underneath the abdomen in an area not easily seen by the beekeeper. Here they force their way between the overlapping segments

and suck the bee's blood. The breeding cycle is also particularly harmful to the bee larvae. Adult varroa move in on unsealed larvae and are subsequently sealed in by the workers. The developing mites feed off the blood of the bee larvae and the female varroa attach themselves to the emerging bees, most of which appear to suffer little from the blood letting. In cases of severe infestation, however, the bee larvae can be killed or the adult bees can emerge greatly deformed. Varroa appear to prefer drone larvae to worker larva.

Male varroa are smaller than the females, are lighter in colour and appear unable to survive outside the sealed brood cells.

Most beekeepers in the British Isles will already be taking some sort of precautions against their bees becoming infested with varroa. A few questions asked of local beekeepers will soon supply information on the likely level of varroa infection in your area. New beekeepers taking up the craft are now in the position of having to consider varroa control as part of their annual colony management.

To ignore varroa is not an option. Colonies with high levels of infestation will eventually perish, often with surprising rapidity. It would appear that the varroa itself is not the killer, but that the varroa spreads latent bee viruses, which actually kill the bees. This condition where colonies fail and die out is known as colony collapse.

The whole subject of varroa detection and control is too extensive to be dealt with here. MAFF publish an excellent booklet giving full details of the mite, its life style, effects on bees, ways of control, and a list of licensed and unlicensed varroacides available at the present time (see page 108).

Varroa, which is generally found on the bee's abdomen (usually on the underside)

Before you have been too long in beekeeping circles you will hear talk of alternative systems; such expressions as 'brood and a half' and 'double-brood chambers' will inevitably crop up. These expressions refer to systems whereby a single brood body has been expanded to offer extra breeding space for very prolific bees by putting a second box for breeding on top. In the case of double brood a second brood body is placed on top of the first and a queen excluder placed on top of both under the supers. This effectively doubles the egg laying area available to the queen.

In brood and a half a second breeding area is added, but this time a super full of frames is given instead of a second brood body. The breeding super can be above or below the brood body, although it is more usual to have it above. A queen excluder is placed on top of the two boxes to stop the queen going up into the honey storing supers.

Both these systems are in fairly common use. I understand that initially they were stop-gap measures adopted by beekeepers using smaller hives of the National or WBC type, who changed to the more prolific Italian-type bees and then discovered the queens were short of laying space. By adding an extra box for breeding one problem was solved, but others were introduced.

When the two boxes are stacked in position there is a gap of one bee space between the bottom of the frames in the top box and the top of the frames in the lower box. If the queen is to use both boxes for breeding, the brood nest will thus be split into two

parts. The bees do not like this and tend to build brace comb in between. This makes it very difficult to remove frames for inspection without stirring up the bees and making them angry.

The second problem comes during the swarming months when the frames have to be inspected regularly by the beekeeper for the appearance of queen cells. Some enthusiasts of double brood and brood and a half systems claim the queen cells always appear between the two boxes, so that if the top box is tipped up at an angle and its underside inspected the queen cells will be visible without the frames being removed. Their bees might work to these rules but mine didn't. I found that I was inspecting twice the number of frames, and soon changed back to a single brood body system as soon as possible. My view is that what started off as a stop-gap became regular practice, without too much thought about the reasons behind it. If the bees are so prolific that more space is required than can be provided with a smaller hive, the logical solution is to change to a larger hive, such as the Langstroth or Modified Commercial. Users of National equipment can effect the change relatively easily by changing to Modified Commercial brood body and frames; the floor, roof, crown board and supers will not need replacing as they are compatible with the new brood body. An alternative solution is to change the bees to a less prolific strain.

Before changing to any larger hive or even considering a double brood body system it is well worth making sure that all the available space in the

Left to right: single brood, brood and a half, double brood

hive can be used by the queen. Every time a cell is used for breeding a thin cocoon skin is left lining the inside. In time these can build up in thickness and so reduce the space available that the cell is no longer of use for breeding. It is possible that the queen will only use the cell with reluctance, even before this stage is reached. For this reason many experts advocate replacing about one quarter of the older brood frames with frames of new foundation each year. If this is done regularly in early spring the queen will usually have the whole of the area of the brood body available for egg laying at the height of the season.

Some experts can quote examples of perfectly normal bees occupying comb known to be fifteen to twenty years old. In such cases the cell walls have been found to be of normal size. It is thought that the bees themselves strip out the cocoon skins from the cell walls when necessary. It does appear, however, that given a choice the queen will lay much more readily on new comb than on old so that regular spring replacement of old comb by new foundation should be considered as a realistic alternative to increasing the size of the brood body.

Cold and warm ways

In hives such as the Langstroth, the frames are arranged so that they run from the entrance, at the front, to the back of the hive. This arrangement is called 'cold way'. Some hives with square brood bodies, such as the National and Modified Commercial, can be placed on the floor so that the frames are arranged in this same 'cold way' or they can be turned through a right angle so that the frames are then arranged from one side to the other when looked at from the front. This is called 'warm way', as it is believed the outside frame nearest the entrance presents a solid block to the wind blowing in through the entrance, and therefore keeps the brood nest warmer.

Beekeepers have varying preferences — some use 'cold way' in summer and 'warm way' in winter. There are advantages and disadvantages claimed for both 'ways'. On balance, there is probably some slight advantage in summer ventilation by using 'cold way', but the differences between the two systems of

Left: cold way; **right:** warm way

manipulation are probably of more interest to their devotees than they are to the bees they manage.

Conclusion

There are as many ways of managing bees as there are of managing a garden. Your bees, hives and equipment will be different from any other beekeeper's, so what works for someone else might not work for you — and vice versa.

Read, observe, listen, and above all discuss your problems and successes. All beekeepers love to talk about their hobby. Find out what works and adopt or adapt it. Don't be afraid to discuss your failures — we all have them.

With a colony of bees at the bottom of the garden not only yours but every garden in the road will benefit from the bees' efforts at pollination. Hopefully, you will have enough honey for your family and some to spare for Christmas presents and bring-and-buys. Do not expect your bees to make a fortune for you overnight. You will only be disappointed.

As a hobby, beekeeping will repay all your expenditure through the honey crop, will introduce you to a mass of other beekeepers and will give you charge of a fascinating colony of industrious and dedicated workers.

Clipping the queen's wings

Many beekeepers clip their queens' wings as an essential part of swarm control procedure. The queen is found, held firmly, and between one-half and one-third of one pair of wings is removed with a pair of small nail scissors, with especial care being taken not to remove any legs as well. The queen is then quietly and carefully returned to the frame she was on.

Clipping a queen's wings in this way alters the normal pattern of behaviour of a colony preparing to swarm. An unclipped queen will normally fly off with the swarm when the first queen cell is sealed, and sometimes even before this. Clipped queens, on

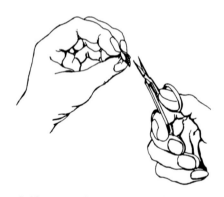

How to hold a queen when clipping one of her wings

the other hand, usually attempt to swarm just before the first virgin queen emerges from her cell, i.e. a week after the cell was sealed. This delay of seven days gives the beekeeper who clips his queens extra breathing space, and he normally cuts down his routine swarm control inspections from every seven days to every ten to fourteen days. In this way he

reduces the number of routine inspections, saves himself time and effort, and disturbs the bees less often.

A clipped queen attempting to fly off with a swarm will be unable to do so and will usually fall down and get lost in the grass near the hive entrance. Should this happen the swarming workers will normally sense her absence, return to the hive and await the emergence of the first virgin queen. The colony might then swarm again, this time with the virgin queen, so clipping a queen's wings is not by itself a sure method of swarm control.

The clipping of queens' wings is an emotive subject. Many beekeepers claim it is cruel to mutilate the queen in this way, that she will be rejected by the colony and quickly replaced. Supporters of queen clipping claim that clipping a queen's wings is no more injurious to the queen than having hair or finger nails cut is to humans.

Personally, I have seen no evidence of queens being immediately replaced and have had several clipped queens laying well two or three years after being clipped. On ther other hand, I have damaged a queen by rough handling during the clipping and marking operations. Both require care and skill, and it is suggested you practise on drones first before attempting to clip a valuable queen.

Discuss the subject of queen clipping with your beekeeping friends and should you decide to go ahead perhaps a more experienced colleague will clip your queen for you.

The subject is highly contentious. I have copied many other beekeepers and clip all my queens as a regular part of swarm control procedure, yet I know of many beekeepers who do not, and some of them would rather tear off their own arms than clip off their queens' wings!

Name	Dimensions of brood body	Number of frames in brood body	Comb area in brood body	Approximate number of worker cells in brood body	Top or bottom bee space	Notes
WBC	18⅛×16½in	10 British Standard	2,000in²	45,000	Bottom	The only double-walled hive
Smith	18¼×16⅜in	11 British Standard (with short lugs)	2,200in²	50,000	Top	Popular in Scotland
National and Modified National	18⅛×18⅛in	11 British Standard	2,200in²	50,000	Bottom	The most popular hive in Britain
Langstroth	20×16¼in	10 self-spacing	2,750in²	62,000	Top	The world's most popular hive
Modified Commercial	18¼×18¼in	12 Self-spacing (but normally 11 + a dummy board)	3,300in² (with 12 frames)	75,000	Bottom	Interchangeable with National except for frames
Modified Dadant	20×18½in	11 self-spacing	3,750in²	85,000	Top	The largest hive in common use

Metric conversion

16¼in – 414mm   16⅜in – 416mm   16½in – 419mm   18⅛in – 460mm   18¼in – 464mm   18½in – 470mm   20in – 508mm

Publications

Magazines and Periodicals

Beecraft *(monthly beekeeping magazine: official journal of the British Beekeepers' Association)*
Secretary
Sue Jakeman
Bee Craft Ltd
107 Church Street
Werrington
Peterborough
PE4 6QF
Email: secretary@bee-craft.com

Beekeepers Quarterly *(an excellent independent publication containing articles on all aspects of beekeeping from contributors from many different countries)*
Northern Bee Books
Scout Bottom Farm,
Mytholmroyd,
Hebden Bridge,
West Yorkshire
HX7 5JS

BBKA Publications
BBKA publish information and advisory leaflets from time to time. A stamped addressed envelope to the BBKA Administrative Headquarters, Stoneleigh and brief note requesting the current availability of advisory leaflets is suggested.

Ministry of Agriculture, Fisheries and Food publications
MAFF publish excellent leaflets on bee diseases and varroa control measures. These publications can usually be obtained from your Bees Officer (also known as Regional and Seasonal Bee Inspectors), or by post from the National Bee Unit, Sand Hutton, York YO4 1LZ.

Books about beekeeping
Details of most of the beekeeping books currently available are given in the catalogues supplied by beekeeping equipment manufacturers (see p. 110). There are also some specialist suppliers dealing in both new and out of print books. One such firm is Northern Bee Books, Scout Bottom Farm, Mytholmroyd, Hebden Bridge, West Yorkshire HX7 5JS.

   Northern Bee Books also publish a Beekeepers' Annual which is packed with information useful to beginner and expert alike. Write for their lists, enclosing a stamped addressed envelope.

Varroa Detection and Control

All beekeepers should take steps to obtain and study the MAFF booklet Varroosis – a parasitic infestation of honey bees. MAFF has been engaged in collecting research information on all aspects of varroa for some years before and after varroa actually appeared in the UK. The process has continued and a great deal of research information is now available. The booklet provides in depth information on all aspects of varroa detection and control.

It has been said one of the main effects of the arrival of varroa in an area is that many, often more senior, beekeepers, give up the craft. I some cases they consider that they were too old to learn the new skills necessary, while in others they could not believe that their bees could become infected and die out, but it happened. In many parts of Europe however the beekeepers that adapted now manage their colonies by controlling varroa levels and many enjoy good yields of honey.

In areas where an infestation is unlikely, beekeepers are encouraged to engage in a series of regular checks, with the aim of discovering an infestation as soon as possible. Some sort of remedial action can then be taken. In areas where varroa is known to exist it is safest to assume that positive actions will need to be taken on a regular basis to protect ones bees.

In practice British beekeepers who are coping with varroa infestation seem to adopt one of two strategies. The first system assumes that varroa are present and active within the hive and involves treating the colony, with a proprietary chemical treatment, in the Autumn after the honey has been removed. This should dramatically reduce the varroa mites present during the Winter, without of course affecting the honeybees. This system, which sometimes involves a second chemical treatment in the Spring, should enable the colony to work throughout the summer reasonably untroubled by their nasty parasites.

The alternative system involves an ongoing programme of checks of the level of infestation, with treatment given when the infestation has reached a particular point. This method of control can involve a non-chemical way of detection which is somewhat gruesome, yet particularly effective. An uncapping fork is used to remove some sealed drone larvae and varroa mites attached to the larvae can easily be spotted, and the level of infestation assessed.

Research into Varroa control is going on throughout Europe, with governments using the findings to adopt national policies. Just as there are many ways of killing a cat, there are many ways of killing varroa, some more effective than others. Some beekeepers prefer chemical treatments while others prefer Biotechnical methods, which do not require the use of chemicals. In UK the MAFF issue licences for certain chemical treatments to be sold. These can be purchased from beekeeping equipment suppliers who advertise their products in the beekeeping press.

As your own varroa treatment will depend on so many factors, and you might chose one of several methods of detection and control, it is difficult to include varroa treatment in the section on the Beekeepers Year (pages 92 & 93). It must again be stressed however, that varroa cannot be ignored so time spent studying the Varroosis booklet or in attending study sessions with your local beekeeping association are of vital importance.

Useful addresses

The following addresses were correct at the time of publication.

BBKA Administrative Headquarters,
National Agricultural Centre,
Stoneleigh,
Kenilworth,
Warwickshire CV8 2LZ.

National Beekeeping Unit (NBU)
Central Science Laboratory,
National Bee Unit,
Sand Hutton,
York,
North Yorkshire YO4 1LZ.

Manufacturers and suppliers of beekeeping equipment

E H Thorne (Beehives) Ltd,
Beehive Works, Wragby,
Lincoln LN3 5LA.

Sherriff International,
Carclew Road,
Mylor, Falmouth TR11 5UN.

KBS,
Brede Valley Bee Farm, Cottage Lane,
Westfield,
Hastings TN35 4RT.

Maisemore Apiaries,
Old Road,
Maisemore,
Gloucestershire GL2 8HT.

Thornes of Scotland
Unit 1, Newburgh Industrial Estate
Cupar Road
Newburgh, Fife
KY14 6HA

Thornes of Stockbridge
Chilbolton Down Farm
Chilbolton Down
Stockbridge
Hampshire
SO20 6BU

Index